FAULT-TOLERANCE TECHNIQUES FOR SRAM-BASED FPGAS

T0137759

FRONTIERS IN ELECTRONIC TESTING

Consulting Editor
Vishwani D. Agrawal

Books in the series:
Data Mining and Diagnosing IC Fails
Huisman, L.M., Vol. 31
ISBN: 0-387-24993-1
Fault Diagnosis of Analog Integrated Circuits
Kabisatpathy, P., Barua, A. (et al.), Vol. 30
ISBN: 0-387-25742-X
Introduction to Advanced System-on-Chip Test Design and Optimi...
Larsson, E., Vol. 29
ISBN: 1-4020-3207-2
Embedded Processor-Based Self-Test
Gizopoulos, D. (et al.), Vol. 28
ISBN: 1-4020-2785-0
Advances in Electronic Testing
Gizopoulos, D. (et al.), Vol. 27
ISBN: 0-387-29408-2
Testing Static Random Access Memories
Hamdioui, S., Vol. 26
ISBN: 1-4020-7752-1
Verification by Error Modeling
Radecka, K. and Zilic, Vol. 25
ISBN: 1-4020-7652-5
Elements of STIL: Principles and Applications of IEEE Std. 1450
Maston, G., Taylor, T. (et al.), Vol. 24
ISBN: 1-4020-7637-1
Fault Injection Techniques and Tools for Embedded systems Reliability ...
Benso, A., Prinetto, P. (Eds.), Vol. 23
ISBN: 1-4020-7589-8
Power-Constrained Testing of VLSI Circuits
Nicolici, N., Al-Hashimi, B.M., Vol. 22B
ISBN: 1-4020-7235-X
High Performance Memory Testing
Adams, R. Dean, Vol. 22A
ISBN: 1-4020-7255-4
SOC (System-on-a-Chip) Testing for Plug and Play Test Automation
Chakrabarty, K. (Ed.), Vol. 21
ISBN: 1-4020-7205-8
Test Resource Partitioning for System-on-a-Chip
Chakrabarty, K., Iyengar & Chandra (et al.), Vol. 20
ISBN: 1-4020-7119-1
A Designers' Guide to Built-in Self-Test
Stroud, C., Vol. 19
ISBN: 1-4020-7050-0
Boundary-Scan Interconnect Diagnosis
de Sousa, J., Cheung, P.Y.K., Vol. 18
ISBN: 0-7923-7314-6
Essentials of Electronic Testing for Digital, Memory, and Mixed Signal VLSI Circuits
Bushnell, M.L., Agrawal, V.D., Vol. 17
ISBN: 0-7923-7991-8
Analog and Mixed-Signal Boundary-Scan: A Guide to the IEEE 1149.4 Test ...
Osseiran, A. (Ed.), Vol. 16
ISBN: 0-7923-8686-8

FAULT-TOLERANCE TECHNIQUES FOR SRAM-BASED FPGAS

by

FERNANDA LIMA KASTENSMIDT
UFRGS, Instituto de Informatica, Porto Alegre, Brazil

LUIGI CARRO
UFRGS, Departamento de Engenharia Eletrica, Porto Alegre, Brazil

and

RICARDO REIS
UFRGS, Instituto de Informatica, Porto Alegre, Brazil

 Springer

A C.I.P. Catalogue record for this book is available from the Library of Congress.

ISBN-10 0-387-31069-X (e-book)

ISBN-13 978-1-4419-4052-0   ISBN-13 978-0-387-31069-5 (e-book)

Published by Springer,
P.O. Box 17, 3300 AA Dordrecht, The Netherlands.

www.springer.com

Printed on acid-free paper

Printed in the Netherlands.

Dedication

To my husband Christopher and to my parents Luiz Fernando and Ana Maria, who always gave me the support to follow my dreams.
"Fernanda Lima Kastensmidt"

*To the future
so that it can be appreciated
by Erika while it arrives
To my parents,
Cesare and Esther
that have educated me
on thinking about it.*
"Luigi Carro"

To my wife Lucia, to my daughter Mariana, to my sons Guilherme and Eduardo, and to my parents Constantino and Maria de Lourdes.
"Ricardo Reis"

Contents

DEDICATION v

AUTHORS ix

PREFACE xi

1. INTRODUCTION 1

2. RADIATION EFFECTS IN INTEGRATED CIRCUITS 9

2.1 RADIATION ENVIRONMENT OVERVIEW 9
2.2 RADIATION EFFECTS IN INTEGRATED CIRCUITS 13
    2.2.1 SEU Classification.. 16
2.3 PECULIAR EFFECTS IN SRAM-BASED FPGAS.. 17

3. SINGLE EVENT UPSET (SEU) MITIGATION TECHNIQUES 29

3.1 DESIGN-BASED TECHNIQUES.. 31
    3.1.1 Detection Techniques.. 32
    3.1.2 Mitigation Techniques 33
        3.1.2.1 Full Time and Hardware Redundancy 33
        3.1.2.2 Error Correction and Detection Codes 39
        3.1.2.3 Hardened Memory Cells.. 43
3.2 EXAMPLES OF SEU MITIGATION TECHNIQUES IN ASICS 54
3.3 EXAMPLES OF SEU MITIGATION TECHNIQUES IN FPGAS........................ 61
    3.3.1 Antifuse based FPGAs 62
    3.3.2 SRAM-based FPGAs.. 65
        3.3.2.1 SEU Mitigation Solution in high-level description 66
        3.3.2.2 SEU Mitigation Solutions at the Architectural level 67
        3.3.2.3 Recovery technique 69

4. ARCHITECTURAL SEU MITIGATION TECHNIQUES.....................73

5. HIGH-LEVEL SEU MITIGATION TECHNIQUES...........................83

5.1 TRIPLE MODULAR REDUNDANCY TECHNIQUE FOR FPGAS 84
5.2 SCRUBBING 88

6. TRIPLE MODULAR REDUNDANCY (TMR) ROBUSTNESS..............91

6.1 TEST DESIGN METHODOLOGY 95
6.2 FAULT INJECTION IN THE FPGA BITSTREAM 96
6.3 LOCATING THE UPSET IN THE DESIGN FLOORPLANNING 98
   6.3.1 Bit column location in the matrix...99
   6.3.2 Bit row location in the matrix 100
   6.3.3 Bit location in the CLB 100
   6.3.4 Bit Classification..101
6.4 FAULT INJECTION RESULTS...103
6.5 THE "GOLDEN" CHIP APPROACH 108

7. DESIGNING AND TESTING A TMR MICRO-CONTROLLER........111

7.1 AREA AND PERFORMANCE RESULTS 114
7.2 TMR 8051 MICRO-CONTROLLER RADIATION GROUND TEST RESULTS ..116

8. REDUCING TMR OVERHEADS: PART I 123

8.1 DUPLICATION WITH COMPARISON COMBINED WITH TIME
   REDUNDANCY..124
8.2 FAULT INJECTION IN THE VHDL DESCRIPTION 132
8.3 AREA AND PERFORMANCE RESULTS 136

9. REDUCING TMR OVERHEADS: PART II...143

9.1 DWC-CED TECHNIQUE IN ARITHMETIC-BASED CIRCUITS 145
   9.1.1 Using CED based on hardware redundancy 148
   9.1.2 Using CED based on time redundancy...150
   9.1.3 Choosing the most appropriated CED block 152
      9.1.3.1 Multipliers..152
      9.1.3.2 Arithmetic and Logic Unit (ALU)....................................153
      9.1.3.3 Digital FIR Filter..154
   9.1.4 Fault Coverage Results..154
   9.1.4 Area and Performance Results 158
9.2 DESIGNING DWC-CED TECHNIQUE IN NON-ARITHMETIC-BASED
   CIRCUITS 168

10. FINAL REMARKS 171

REFERENCES 175

Authors

Fernanda Gusmão de Lima Kastensmidt is a professor of Computer Science at the Federal University of Rio Grande do Sul (UFRGS) located in Porto Alegre, Brazil. She received her BS in Electrical Engineering in 1997 and MS and PhD degrees in Computer Science and Microelectronics in 1999 and 2003, respectively, from the Federal University of Rio Grande do Sul (UFRGS), Porto Alegre, Brazil. She has worked in the Grenoble National Polytechnic Institute (INPG), France, in 1999 and Xilinx Corporation, San Jose, USA, in 2001. Her research interests include VLSI testing and design, fault effects, fault tolerant techniques and programmable architectures. She is an IEEE member.

Luigi Carro was born in Porto Alegre, Brazil, in 1962. He received the Electrical Engineering and the MSc. degrees from Universidade Federal do Rio Grande do Sul (UFRGS), Brazil, in 1985 and 1989, respectively. From 1989 to 1991 he worked at ST-Microelectronics, Agrate, Italy, in the R\&D group. In 1996 he received the PhD. degree in the area of Computer Science from Universidade Federal do Rio Grande do Sul (UFRGS), Brazil. He is presently a lecturer at the Electrical Engineering Department of UFRGS, in charge of Digital Systems Design and Digital Signal processing disciplines at the graduate and undergraduate level. He is also a member of the Graduation Program in Computer Science of UFRGS, where he is responsible for courses in Embedded Systems, Digital Signal Processing, and VLSI Design. His primary research interests include mixed-signal design, digital signal processing, mixed-signal and analog testing, and fast system prototyping. He has published more than 90 technical papers in those topics and is the author of the book Digital Systems Design and Prototyping (in portuguese).

Ricardo Reis is a professor at the Instituto de Informatica of the Universidade Federal do Rio Grande do Sul (UFRGS), Brazil. He received the Electrical Engineering degree from the Federal University of Rio Grande do Sul (UFRGS), Porto Alegre, Brazil, in 1978. In 1983 he received the PhD. degree in the area of Computer Science and Microelectronics from the Institut National Polytechnique de Grenoble (INPG), France. His primary research interests include VLSI design and CAD, physical design, design methodologies, fault tolerant techniques. He has published more than 200 technical papers in journals and conferences as well published some books. He is a former president of the Brazilian Computer Society and former vice-president of the Brazilian Microelectronics Society. He is a vice-president of the International Federation for Information Processing, IFIP. He received the Silver Core from IFIP. He is the Editor-in-Chief of the Journal of Integrated Circuits and Systems, JICS. Ricardo is also Latin America liaison of the IEEE D&T. He contributed to the organizing and program committees of several conferences and he is a founder of the SBCCI conference series (Symposium on Integrated Circuits and Systems Design). He is a member of IEEE.

Authors

Preface

This book presents fault-tolerant techniques for programmable architectures, the well-known Field Programmable Gate Arrays (FPGAs), customizable by SRAM. FPGAs are becoming more valuable for space applications because of their high density, high performance, reduced development cost and re-programmability. In particular, SRAM-based FPGAs are very valuable for remote missions because of the possibility of being reprogrammed by the user as many times as necessary in a very short period. SRAM-based FPGA and micro-controllers represent a wide range of components in space applications, and will be the focus of this work, more specifically the Virtex® family from Xilinx and the architecture of the 8051 micro-controller from Intel.

The content addressed in this book ranges from the study of state-of-the-art of SEU mitigation techniques for ASIC and FPGA components, to the implementation and test of new fault-tolerant techniques for SRAM-based FPGA components. In the first phase of the research presented in this text, available techniques to protect integrated circuits against radiation are studied. The radiation fault-tolerant techniques can be classified as: the ones that change the technology used in the fabrication process such as Silicon on Insulator (SOI), and the ones that change the hardware design of a system such as SEU hardened memory cells, error detection and correction codes (EDAC) and logic redundancy. There is a trade-off with each mitigation technique for each type of architecture system, and there is no best unique solution so far. Some of the considered techniques are evaluated in terms of area, cost and performance. The first case study circuit is the 8051 micro-controller from Intel (Intel, 1994). The microprocessor architecture was chosen for its representation of the majority of system requirements in space applications nowadays, presenting all types of logic to be protected and being part of the new generation architectures based on FPGA with an embedded hard microprocessor core.

The description of the 8051 micro-controller used in the experiment was developed at Federal University of Rio Grande do Sul (Carro; Pereira; Suzim, 1996). It is composed of a datapath unit, control unit, state machine, instruction decoding unit and embedded memory. Although the 8051 micro-controller has a simplified architecture compared to the latest available microprocessors, the assumptions made in its architecture can be adapted to any other processor-like circuit. Techniques such as hamming code and radiation tolerant flip-flops were implemented in the 8051 micro-controller (LIMA et al., 2000a; LIMA et al., 2000b). Fault injection (LIMA et al., 2001a; LIMA et al., 2002a; LIMA et al., 2002b) and simulation are used to analyze the efficiency of the techniques. Area and performance are taken into consideration with the results.

The second phase of the work presented in this book resumes the analysis of an SRAM-based FPGA and the SEU effects in this architecture. The Virtex® FPGA family from Xilinx is the most popular high density and high performance FPGA used in the market nowadays, and it was chosen to be the object of study in this work. There are two ways to mitigate SEU in FPGA designs, as previously mentioned. One is based in changing the FPGA architecture and the other one is based on modifying the high-level design description before the FPGA synthesis. First, implementations of some SEU mitigation techniques in the architectural level of the FPGA matrix are proposed. The SRAM-based architecture is divided in main blocks classified by functionality (such as LUT), flip-flops, customization routing, embedded memory, PLL, etc. SEU mitigation techniques for many of the blocks are discussed. The objective is to show the trade-off of each technique in the Virtex® FPGA and the complexity of developing a new architecture with changes at the mask level. This investigation is based on the experience collected in first phase.

Because of the limitations in developing and testing a new fault-tolerant FPGA architecture such as cost and time-to-market, techniques at the high-level description must also be investigated. The Triple Modular Redundancy (TMR) with voters is a common technique to protect against SEU in ASICs and it can be also applied to protect FPGAs against SEU, as shown in (Carmichael, 2001). In this case, the mitigation can be applied to the high-level design description language and synthesized in the device without any changes in the mask process. The TMR technique is first tested in the Virtex® architecture by using a small design based on counters. Faults are injected in all sensitive parts of the FPGA and a detailed analysis of the effect of a fault in a TMR design synthesized in the Virtex® platform is performed.

In order to test a more complex design protected by TMR in the Virtex® platform that would also include embedded memories, the same 8051 like micro-controller description was protected by TMR and tested under the FPGA platform. There are many advantages of using the same design as the 8051 micro-controller, such as good description knowledge, importance of micro-controllers IP in FPGA and the possibility of comparison with the previous techniques (hamming code and SEU hardened memory cells) applied in the same description. The TMR 8051 micro-controller was tested by fault injection and under proton radiation in a ground facility (Lima et al., 2001b). At the end of these practical experiments (Lima et al., 2001b; Carmichael; Fuller; Fabula; Lima, 2001), the use of TMR in Virtex® FPGAs has confirmed the efficacy of the TMR structure to recover upsets in the FPGA architecture. However, the TMR technique presents some limitations, such as area overhead, three times more input and output pins and, consequently, a significant increase in power dissipation and also some robustness issues. The result has brought the necessity of improving this technique in order to reduce the overheads and to try to improve robustness as well.

In the third phase of the work, additional SEU mitigation techniques for the Virtex® FPGA architecture are investigated. A new high-level fault-tolerant technique for SRAM-based FPGA was developed (Lima, Carro, Reis, 2003a; Lima, Carro, Reis, 2003b). This technique combines time and hardware redundancy with some extra features able to cope with the effects of SEU in FPGAs, and at the same time it is able to reduce the number of input and output pads and area overhead compared to the traditional TMR approach. The methodology was validated in combinational and sequential circuits by using fault injection experiments emulated in a prototype board. Results have confirmed that this new technique can reduce not only pin count but also area as well, without compromising performance and reliability.

This book is organized as follows. Chapter 2 describes the radiation effects on integrated circuits manufactured using CMOS process and it explains in detail the difference between the effects of a SEU in ASIC and in SRAM-based FPGA architectures. This chapter shows the architecture analysis of the Virtex® FPGA and all its radiation sensitive area. Chapter 3 presents the main techniques, either being commercialized by companies or being studied by researchers, to mitigate the effects of radiation in ASICs, such as microprocessors and memories, and in programmable architectures, such as FPGAs programmed by SRAM and by antifuse technology.

Chapter 4 discusses some SEU mitigation techniques that can be applied at the FPGA architectural level. The FPGA is divided by functionality in main logic blocks. Each block has different characteristics, and the fault-tolerant technique must take into account the peculiarities of each. In the end, a SEU tolerant FPGA is proposed based on the presented SEU mitigation techniques.

Chapter 5 defines the problem of protecting SRAM-based FPGAs against radiation in the high level description. The Triple Modular Redundancy (TMR) technique in the high level description for FPGAs is addressed in this chapter. Chapter 6 evaluates the robustness of the TMR technique by using fault injection in the bitstream of the FPGA and also in a radiation ground test facility. In this chapter, a methodology is presented to relate the upset bit in the bitstream to the SRAM cell location in the user's design floor-planning. The obtained results represent an important base for this work, because it shows the limitations of the TMR method on the SRAM-based FPGA, justifying the research of new design techniques for SEU mitigation in SRAM based FPGAs.

Chapter 7 shows the implementation and results of the 8051 description protected by TMR in the Virtex® FPGA. All implementation details of the TMR technique were carefully applied to the VHDL description of the 8051 to test this technique in a more complex design. The final protected design was tested by fault injection in the FPGA bitstream and also in a radiation ground test facility. Results and final remarks are placed at the end of that chapter.

Chapter 8 introduces a new high-level technique for designing fault tolerant systems for SRAM-based FPGAs, without modifications in the FPGA architecture, able to cope with transient faults in the user combinational and sequential logic, while also reducing pin count, area and power dissipation compared to the traditional TMR. The methodology is validated by fault injection experiments in VHDL description emulated in a prototyped board. Results in terms of fault coverage and area and performance comparison with the TMR approach are presented.

The technique presented in chapter 8 presents some limitations in fault coverage because it uses the standard time redundancy approach to detect the effect of a SEU in the FPGA matrix. In chapter 9, an improvement to the high-level technique presented in chapter 8 is proposed. This technique combines duplication with comparison and concurrent error detection technique in order to cope with the permanent effects of a SEU in FPGAs and at the same time to reduce TMR overheads. In addition, this proposed method is also able to detect physical faults, which are permanent faults that are not corrected by reconfiguration. The methodology is also validated by fault injection experiments in an emulation board. Some fault coverage results and a comparison with the TMR approach are evaluated.

Final remarks are placed in chapter 10 followed by the references. Because the technology is constantly in evolution, there are always improvements to be made in the protection of integrated circuits, and consequently, in the way designs are protected against faults. This work has contributed to some solutions for the SRAM-based FPGAs that are being projected to work in commercial applications but are manufactured by nanotechnologies and need to work properly in the presence of upsets. However, there is much more research to be done as each step of investigation brings more questions and possibilities of solutions. As a result, future techniques are discussed in the final chapter of this book.

Chapter 1

INTRODUCTION
Fault Tolerance in SRAM-based FPGAs

Fault tolerance on semiconductor devices has been a meaningful matter since upsets were first experienced in space applications several years ago. Since then, the interest in studying fault-tolerant techniques in order to keep integrated circuits (ICs) operational in such hostile environment has increased, driven by all possible applications of radiation tolerant circuits, such as space missions, satellites, high-energy physics experiments and others (Nasa, 2003).

Spacecraft systems include a large variety of analog and digital components that are potentially sensitive to radiation and must be protected or at least qualified for space operation. Designers for space applications currently use radiation-hardened devices to cope with radiation effects. However, there is a strong drive to utilize standard commercial-off-the-shelf (COTS) and military devices in spaceflight systems to minimize cost and development time as compared to radiation-hardened devices (Katz et al., 1997; Obryan and Label, 2001).

The space radiation environment can have serious effects on spacecraft electronics. Single Event Effect (SEE) is the main concern in space (Barth, 1997), with potentially serious consequences for the application, including loss of information and functional failure. SEE occurs when charged particles hit the silicon transferring enough energy in order to provoke a fault in the system.

SEE can have a destructive or transient effect, according to the amount of energy deposited by the charged particles and the location of the strike in the device. The main consequences of the transient effect, also called Single Event Upset (SEU), are bit flips in the memory elements. SEU has been constantly magnified in the past years, caused by the continuous technology evolution that has led to more and more complex architectures, with a large amount of embedded memories, followed by an amazing scaling down process of transistor dimensions following Moore's Law (Moore, 1975).

The fabrication technology process of semiconductor components is in continuous evolution in terms of transistor geometry shrinking, power supply, speed, and logic density (SIA Semiconductor, 1994). As stated in (Johnston, 2000; Obryan and Label, 2001; Obryan et al., 2002; Dupont et al., 2002), drastic device shrinking, power supply reduction, and increasing operating speeds significantly reduce the noise margins, and thus the reliability that very deep submicron (VDSM) ICs face from the various internal sources of noise. Reliability is the probability of no failure in a given operating period. It is used to measure how good a system is and how frequently it goes down.

The fabrication process is now approaching a point where it will be unfeasible to produce ICs that are free from upset effects. A more significant problem is related to SEU. It is predicted that neutrons produced by sun activity will dramatically affect the operation of future ICs. At the sea level, the energy of these particles is not strong enough to drastically affect the operation of current ICs. But as one approaches 0.1μm, or very low supply voltages, the rate of random errors induced by cosmic neutrons will be unacceptable. The situation is worse at flight altitudes. Alpha particles produced by packaging material are becoming another cause of increasing soft error rates in these technologies (Normand, 2001).

The necessity to protect integrated circuits against upsets has become more and more eminent (Johnston, 2000; Label et al., 2000). Experiments presented in (Normand and Baker, 1993; Normand, 1996, 2001) indicate that neutron particles present in the atmosphere are capable of producing SEE in avionics. Recent studies also show that memory cells composed of transistors with channel length smaller than 0.25 μm and combinational logic composed of transistors with length smaller than 0.13 μm may be subject to upsets while operating in the space environment, or inside the atmosphere (Baumann, 2001; Borel et al., 2001). Remember that current IC are been fabricated in 90 nm CMOS technology and below. Terrestrial applications that are determined as critical such as bank servers, telecommunication servers and avionics are more and more considering the use of fault-tolerant techniques to ensure reliability.

Both discussed factors, the space market interest of using COTS/military devices in space applications and the constant increase in the radiation sensitivity of integrated circuits driven by the process scaling, have brought the necessity of researching fault-tolerant techniques for ICs able to cope with the radiation effects at sea level, and also qualifying the design for space applications. Based on the definition of fault-tolerance, the goal is to maintain the IC operating correctly despite the existence of upsets. Although many techniques have been developed in the last few years attempting to avoid SEU, efficient fault-tolerant solutions are still a challenge for the next generation semiconductor industry, especially because of the complexity of the new architectures.

The development of fault-tolerant techniques is strongly associated with the target device, and it requires a detailed analysis of the effects of an upset on the related architecture. For each type of circuit, there is a set of most suitable solutions to be applied. In the past years, the integrated circuit industry has designed more and more complex architectures in order to improve performance, to increase logic density and to reduce cost. Examples of this development include Application Specific Integrated Circuits (ASICs), microprocessors composed of millions of transistors, high-density Field Programmable Gate Array (FPGA) components and, more recently, System-on-a-Chip (SOC) composed of embedded microprocessors, memories and analog blocks. These architectures have made a dramatic impact on the way systems are designed, providing a large amount of information processing on a single chip. They cover a wide range of applications, from portable systems to dedicated embedded control units and computers. In particular, FPGAs have made a major improvement in systems design by adding the reconfigurability feature, which reduces the time to market and increases the design flexibility.

A set of events have contributed to increase the market for FPGAs figure 1-1 in the last years. The constant advances in technology over the last years, following Moore's law, the gap between FPGAs and ASICs in terms of performance has been reduced to a negligible level for the majority of applications. In addition, the cost for low and medium volumes of FPGA parts has reduced compared to low and medium volumes parts of ASICs. The time to market, the necessity to speed up the design process and prototyping, has also increased the need of using programmable logic.

In the 70s, a system was basically composed of a microprocessor component, a memory chip and discrete logic. In the 80s, a large part of the discrete logic was replaced by ASICs and some part by programmable logic components (FPGA). In the 90s, the discrete logic completely disappeared and the system was composed of microprocessors, memory, ASICs and FPGA components. ASICs are progressively being replaced by FPGAs in many systems as illustrated. In addition, more complex structures are constantly being added to FPGA architectures, supported by substantial increases in logic density and performance in the last few years. Nowadays, FPGAs are also replacing microprocessors and memories as these parts are being added to the FPGA matrix.

Consequently, next generation of FPGA architectures do not claim to reduce that gap between ASIC and programmable logic anymore, but to merge microprocessors and reconfigurability features in the same component in order to improve performance and flexibility (DAC, 2001). FPGAs already provide reconfigurability and high performance for many appli-cations, but the necessity of adding either more performance for applications such as Digital Signal Processing (DSP), using high-bandwidth and reducing the board space, power and cost, has increased the interest of embedding microprocessors in the programmable matrix, as illustrated in more detail in figure 1-2. This experience had started with the soft cores synthesized in the FPGA architecture in order to get the highest performance and density tradeoff (Xilinx, 2000; Altera, 2001). It has arrived at the next level of performance and complexity with the Virtex® II–Pro generation, which has up to four hard core PowerPC core microprocessors from IBM embedded in the matrix (Xilinx, 2001a).

As a consequence, FPGAs are increasingly demanded by spacecraft electronic designers because of their high flexibility in achieving multiple requirements such as high performance, low NRE (Non-Recurring Engineering) cost and fast turnaround time. There are many types of customization in the FPGAs. One of the most popular ones uses SRAM memory cells to customize the FPGA, which makes possible in-the-field customization as many times as necessary in a very short period of time. Examples are the families Virtex®, Virtex®-E and Virtex®-II fabricated by Xilinx. As a result, SRAM-based FPGAs are even more valuable for remote missions by offering the additional benefits of allowing in-orbit design changes, with the aim of reducing the mission cost by correcting errors or improving system performance after launch.

The advantages of using SRAM-based FPGAs for space applications and the increase of logic complexity of the programmable logic with more and more embedded memories and specific architectures such as micro-processors brings us the necessity of researching new SEU mitigation techniques specific for programmable architecture. This book presents the study and development of SEU mitigation techniques for programmable logic architectures, more specifically for SRAM-based FPGAs. The consi-deration of using FPGA for space applications is fairly recent, and there is a lot of work to be done in this area. Presently, there is no efficient solution for SRAM based FPGAs that can ensure 100% reliability in all conditions against SEU.

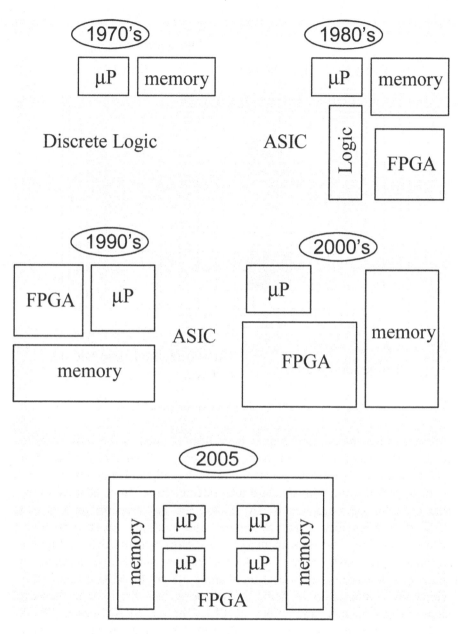

Figure 1-1. Design Evolution using FPGA

Dedicated routing

Ultra fast I/O   Hard IP core   Soft IP core

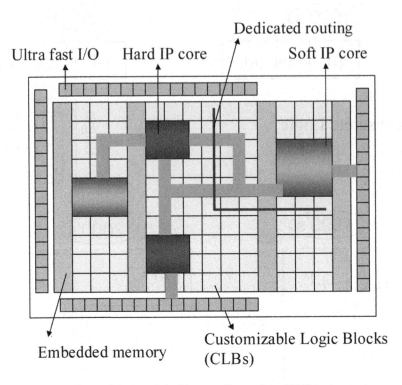

Customizable Logic Blocks
Embedded memory   (CLBs)

Figure 1-2. Actual Architecture Generation of FPGAs

Several fault-tolerant techniques have been studied in the past years to protect ASICs against transient faults, and because FPGAs are composed of combinational and sequential logic, and more recently embedded processors, previous work dealing with standard integrated circuits can be adapted to the programmable logic architecture by finding the best tradeoff among area overhead, performance penalties, single and multiple upset correction, process technology and implementation cost. However, the SEU mitigation techniques previously used for ASICs cannot simply be applied to programmable circuits because of the distinct effect of a SEU in the FPGA architecture compared to an ASIC, as will be further discussed in the next chapter. Consequently, the effect of SEUs in the SRAM-based FPGA architecture must be investigated to identify the limitations of the already used fault-tolerant techniques and to guide the investigation to new solutions.

The goal of this book is to investigate the techniques used nowadays and to present new SEU mitigation techniques for SRAM-based FPGAs that are cost efficient in terms of:

- time to market,

- low development cost,

- high performance,

- low area cost,

- low power dissipation,

- high reliability.

There are two ways to implement fault-tolerant circuits in SRAM-based FPGAs, as exemplified in the flowchart in figure 1-3. The first possibility is to design a new FPGA matrix composed of fault-tolerant elements. These new elements can replace the old ones in the same architecture topology or a new architecture can be developed in order to improve robustness. The cost of these two approaches is high and it can differ according to the development time, number of engineers required to perform the task and the foundry technology used. Another possibility is to protect the high-level description by using some sort of redundancy, targeting the FPGA architecture. In this way, it is possible to use a commercial FPGA part to implement the design and the SEU mitigation technique is applied to the design description before the description is synthesized in the FPGA. The cost of this approach is inferior to the previous one because, in this case, the user is responsible for protecting his/her own design, and the solution does not require new chip development and fabrication. In this way, the user has the flexibility of choosing the fault-tolerant technique and consequently, the overheads in terms of area, performance and power dissipation.

In summary of figure 1-3, the four different implementations of a fault-tolerant FPGA, respectively, A, B, C and D have different costs. Cost B is higher than cost A, which is much higher than cost C, which is also higher than cost D. All of them have their own space in the market, as each application requires different constraints. But since the semiconductor industry tends to emphasize time-to-market and low-cost production, the implementations C and D look more interesting. In this work, both architectural and the high-level methods are presented and discussed, but because of the high cost of the implementations A and B, only implementations C and D are designed and tested in details. Next following chapters present some works that have been developed in these four alternatives solutions to protect SRAM-based FPGAs against SEU.

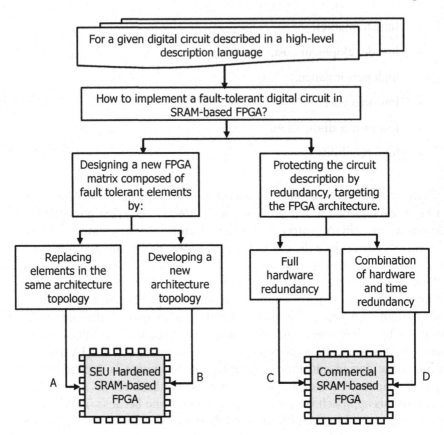

Figure 1-3. Design flow of how to protect a digital circuit implemented in a SRAM-based FPGA, where the cost of solution B is higher than the cost of solution A, which is much higher than cost of solutions C and D

Chapter 2

RADIATION EFFECTS IN INTEGRATED CIRCUITS
Overview

Signal integrity is becoming much more critical in integrated circuits (ICs) designed in very deep sub-micron technologies (VDSM), as device dimensions continue to shrink. Some of the causes are cross coupling and ground bounce, which are increasing the sensitivity of VDSM designs to transient errors (Irom et al., 2002). In addition, ICs operating in space environment and, more recently, at sea level can be upset by charged particles that also generate transient errors in the system. Transient errors provoked by radiation effects are a major concern, and they must be tolerated in order to ensure reliability.

2.1 RADIATION ENVIRONMENT OVERVIEW

The radiation environment is composed of various particles generated by sun activity (Stassinopoulos and Raymond, 1988; BARTH, 1997; Baumann, 2001; Leray, 2001). The particles can be classified as two major types: (1) charged particles such as electrons, protons and heavy ions, and (2) electromagnetic radiation (photons), which can be x-ray, gamma ray, or ultraviolet light. The main sources of charged particles that contribute to radiation effects are protons and electrons trapped in the Van Allen belts, heavy ions trapped in the magnetosphere, galactic cosmic rays and solar flares. The charged particles interact with the silicon atoms causing excitation and ionization of atomic electrons (Obryan et al., 1998).

When a single heavy ion strikes the silicon, it loses its energy via the production of free electron-hole pairs, resulting in a dense ionized track in the local region, as illustrated in figure 2-1 (a). Protons and neutrons can cause nuclear reaction when passing through the material, as illustrated in figure 2-1 (b). The recoil also produces ionization. The ionization generates a charge deposition that can be modeled by transient current pulse that can be interpreted as a signal in the circuit causing an upset.

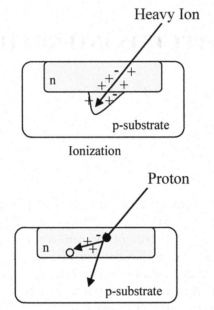

Heavy Ion

Ionization

Proton

Nuclear reaction → Short range recoil → Ionization

Figure 2-1. Charged particle striking the silicon surface

Different radiation sources show different charge deposition waveforms. Moreover, the waveform also depends on the site and angle of incidence, as well as on technological parameters like device doping profile. The charge deposition mechanism is usually modeled by a double exponential current pulse at the particle strike site (MESSENGER, 1982):

$$I_p(t) = I_0 (e^{-t/\tau_\alpha} - e^{-t/\tau_\beta})$$ (1)

Where I_0 is approximately the maximum charge collection current, τ_α is the collection time constant of the junction and τ_β is the time constant for initially establishing the ion track. Figure 2-2 exemplifies a common shape of a transient current pulse induced by a particle hit, note that τ_α is much higher than τ_β. The area under the curve corresponds to the amount of collected charge. The waveform shape can be the decisive factor for a SEU occurrence or not in integrated circuits, as discussed in (WIRTH, et al., 2005a, 2005b).

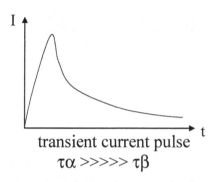

Figure 2-2. Transient current pulse induced by a charged particle hit

The influence of radiation in the material is measured by the energy and the flux of particles. The flux is the number of particles passing during one second through one cm^2 of area $[1/s.cm^2]$. By integrating the flux over time one gets the fluence, which is $[1/cm^2]$. The flux of these sources is affected by the activity of the sun. The energy deposited by the charged particle is measured in rad (1 rad $= 10^{-2}$ Js^{-1}), which corresponds roughly to the generation of $4x10^{13}$ electron-hole pairs in one cm^3 of silicon. The rate at which the particle loses energy is called stopping power (dE/dx). The incremental energy dE is usually measured in units of MeV, while the material thickness is usually measured as a mass thickness in units of mg/cm^2. The energy transferred to the device is called Linear Energy Transfer (LET), and it is measured by the incremental energy per unit length ($MeV/(mg/cm^2)$). The minimum LET that can cause an SEU is called the LET threshold (LET_{th}) (Dentan, 2000). There are many levels of robustness, according to the amount of flux and energy transferred to the silicon that can keep the circuit operating properly.

By counting the number of upsets and knowing how many particles passed through the part, one can calculate the probability of a particular particle causing an upset. This resultant number, which is the number of upsets divided by the number of particles per cm^2 causing the upsets, is called the cross-section of the part and is measured in units of cm^2 / device. Consequently, the sensitivity of a device to an upset is measured by a function of the cross-section (σ) in terms of the LET (Linear Energy Transfer). Figure 2-3 shows an example of cross-section per LET curve.

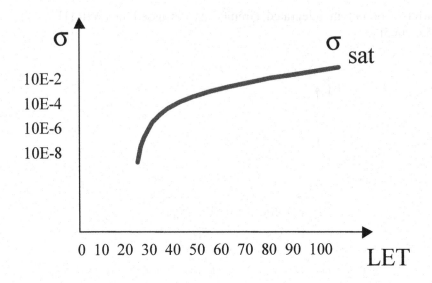

Figure 2-3. An example of cross-section per LET curve

Analyzing the curve of figure 2-3, one can say that no error occurs in the presence of particles with LET (linear energy transfer) lower than 25 MeV. For particles with 25 MeV, more than 100.000.000 particles must travel through the circuit sensitive area to trigger one upset. For particles with 50 MeV, 10.000 particles per second are needed to trigger one upset. And a flux of 100 particles per second with a LET of 100 MeV is needed to trigger one upset.

At the ground level, neutrons are the most frequent cause of upset (Normand, 1996; Obryan et al., 1998; Baumann and Smith, 2000). Neutrons are created by cosmic ion interactions with the oxygen and nitrogen in the upper atmosphere. The neutron flux is strongly dependent on key parameters such as altitude, latitude and longitude. Its peak is around 60,000 ft (~20,000 m). At 30,000 ft (~10,000 m) the neutron flux is about 1/3 of the peak, and on the ground, the flux is about ~1/400 of the peak. At airplane altitudes, the neutron flux is 7,200 neutrons/cm^2/hour. The average neutron flux at sea level is around 20 neutrons/cm^2/hour, however the peak can reach 14,400 neutrons/cm^2/hour.

There are high-energy neutrons that interact with the material generating free electron hole pairs and low energy neutrons. These neutrons interact with a certain type of Boron present in semiconductor material creating others particles, such as Lithium-7, Gamma and Alpha particles (Obryan et al., 1998). The energized alpha particles are the greatest concern in this case, and they are addressable through processing and packaging material. In principle, a very careful selection of materials can minimize alpha particles. However, this solution is very expensive and never eliminates the problem completely (Dupont et al., 2002).

The detailed analysis of the effects of radiation particles in the bulk of a semiconductor is still a challenge. One of the difficulties resides in predicting just what percentage of electron hole pairs is actually collected in the area around the stored data. It is this percentage that determines the critical point at which the radiation induced charge provokes an error in the stored data. Solutions to help the analysis can be the use of complex 3D simulations to help find an accurate shape for the pulse generated by the strike, and the exploration of how the electron-hole-pair cloud can neutralize the stored data.

2.2 RADIATION EFFECTS IN INTEGRATED CIRCUITS

A single particle can hit either the combinational logic or the sequential logic in the silicon (Crain et al., 2001; Alexandrescu et al., 2002). Figure 2-4 illustrates a typical circuit topology found in nearly all sequential circuits. The data from the first latch is typically released to the combinatorial logic on a falling or rising clock edge, at which time logic operations are performed. The output of the combinatorial logic reaches the second latch sometime before the next falling or rising clock edge. At this clock edge, whatever data happens to be present at its input (and meeting the setup and hold times) is stored within the latch.

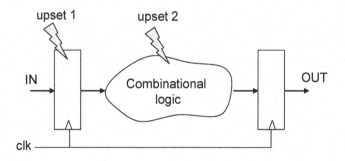

Figure 2-4. Upsets hitting combination and sequential logic

When a charged particle strikes one of the sensitive nodes of a memory cell, such as a drain in an off state transistor, it generates a transient current pulse that can turn on the gate of the opposite transistor. The effect can produce an inversion in the stored value, in other words, a bit flip in the memory cell. Memory cells have two stable states, one that represents a stored '0' and one that represents a stored '1'. In each state, two transistors are turned on and two are turned off (SEU target drains). A bit-flip in the memory element occurs when an energetic particle causes the state of the transistors in the circuit to reverse, as illustrated in figure 2-5. This effect is called Single Event Upset (SEU), and it is one the major concerns in digital circuits.

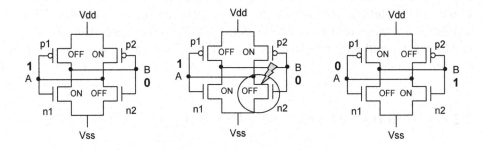

Figure 2-5. Single Event Upset (SEU) effect in a SRAM Memory cell

When a charged particle hits the combinational logic block, it also generates a transient current pulse. This phenomenon is called single transient effect (SET) (Leavy et al., 1991). If the logic is fast enough to propagate the induced transient pulse, then the SET will eventually appear at the input of the second latch in figure 2-4, where it may be interpreted as a valid signal. Whether or not the SET gets stored as real data depends on the temporal relationship between its arrival time and the falling or rising edge of the clock.

Figure 2-6 exemplifies the signal paths in a combinational logic. In (Hass et al., 1998, 1999) the probability of a SET becoming a SEU is discussed. The analysis of SET is very complex in large circuits composed of many paths. Techniques such as timing analysis could be applied to analyze the probability of a SEU in the combinational logic being stored by a memory cell or resulting in an error in the design operation, as presented in (Mohanram, 2005). Additional invalid transients pulses can occur at the combinatorial logic outputs as a result of SETs generated within global signal lines that control the function of the logic. An example of this would be SETs generated in the instruction lines to an ALU (Arithmetic Logic Unit). In (Nicolaidis and Perez, 2003), the widths of some induced transient pulses are measured to obtain more precise models for fault-tolerant analysis.

Please note that according to the logic fan-out, a single SET can produce multiple transient current pulses at the output. Consequently, SETs in the logic can also provoke multiple bit upsets (MBU) in the registers once the SETs are captured by the flip-flops.

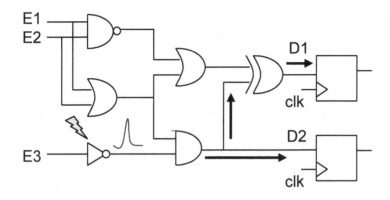

Figure 2-6. Single Event Transient (SET) Effect in Combinational Logic based on (ANGHEL et al., 2000)

Performing a more detailed analysis, the sensitive regions of an integrated circuit are the surroundings of the reverse-biased drain junctions of a transistor biased in the off state (DODD, MASSENGILL, 2003), as for instance the drain of the off p-channel transistor, see figure 2-7. As current flows through the struck transistor, the transistor in the on-state (n-channel transistor in figure 2-7) conducts a current that attempts to balance the current induced by the particle strike. Actually, there are three current components at the struck node. The current induced by the particle strike I_p, the current I_{ON} that flows through the transistor in the on-state, and the current I_C that charges the parasitic capacitances at the node. The current I_C (t) is the current that will charge the node equivalent capacitance and cause the bit flip, and is given by:

$$I_C(t) = I_P(t) - I_{ON}(t) \qquad (2)$$

If the current induced by the particle strike is high enough, the on-transistor can not balance the current and a voltage change at the node will occur. This voltage change can be propagated to the opposite inverter and lead to the flipping of the bit stored in the memory cell. If the voltage transient is feedback through the opposite inverter a SEU occurs. If the voltage on the struck node is recovered by the current feed through the on-transistor no SEU will be observed.

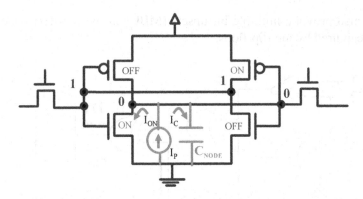

Figure 2-7. Single Event Upset (SEU) effect in a SRAM Memory cell

The critical charge has been reduced in new process technologies because of scaling. For constant field scaling, for example, as all physical device dimensions such as gate length L, gate width W, and gate oxide thickness T_{OX}, are reduced, the supply voltage V_{DD} and the threshold voltage V_{TH} are also reduced proportionately. This fact results in proportionately lower drain current (I_{ON}), proportionately lower load capacitance (C), and proportionately lower circuit gate delay ($C*V_{DD}/I_{ON}$). This means that less charge or current is required to store information. Consequently, devices are becoming more vulnerable to radiation and this means that particles with small charge, which were once negligible, are now much more likely to produce upset.

2.2.1 SEU Classification

SEUs can be classified in first, second and third order effects, according to the number of upsets that occur at the same time in the circuit. A single bit upset (SEU) is classified as a first order effect, while multiple bit upsets (MBU) are classified as second or third order effects. MBU can occur when a single charged particle traveling through the IC at a shallow angle, nearly parallel the surface of the die, simultaneously strikes two sensitive junctions by direct ionization or nuclear recoil (Zoutendyk et al., 1989).

In (Reed et al., 1997), experiments in memories under proton and heavy ions fluxes have shown multiple upsets provoked by a single ion. MBUs were observed for all angles of incidence for LET greater than 25 MeV/(mg/cm^2). There are three types of MBU. The first one occurs when a single particle hits two adjacent sensitive nodes, located in two distinct memory cells. This event is classified as a second-order effect. This type of MBU can be avoided by specific placement, for instance, memory cells of a same register or memory data can be placed far away from each other to avoid the same charged particle stike affect two or more adjacent cell of a same data structure.

The second type of MBU occurs when a single particle strikes two adjacent sensitive nodes located in the same memory cell. This event is classified as a third-order effect. The probability of such a multiple node strike can be minimized in a circuit design by taking care in the physical layout to separate critical node junctions by large distances, and by aligning such junctions so that the area of each junction, as viewed from the other, is minimized.

The third type of MBU occurs when multiple particle strike multiple sensitive nodes in the silicon provoking upsets in multiple memory cells. This event can be analyzed like a group of SEU and it will represent the same immunity characteristics. Based on (Reed et al., 1997), the majority of multiple upsets located in adjacent cells are provoked by a single particle. There is a very low probability of more than one charged particle interacting in adjacent cells, provoking upsets in a period smaller than one second. This can be observed in (Velazco et al., 1999), where it is shown some SEU flight results of two SRAM memories (Hitachi and MHS). A total of 691 upsets were detected for the analyzed period of time, 333 of them arising on the Hitachi SRAM and 358 occurring in the MHS SRAM memory. From this amount only few were multiple upsets, 8 double upsets in the Hitachi and 3 in the MHS memory. The distribution of bit flips within the memory word bits was uniform, and transitions 1 to 0 seem to be slightly more frequent than 0 to 1 for all the tested memories too.

2.3 PECULIAR EFFECTS IN SRAM-BASED FPGAS

The Virtex® family from Xilinx (Xilinx, 2000) is one of the most popular SRAM-based programmable devices used in the market nowadays, because of its high density and high-performance. It supports a wide range of configurable gates, from 50k to more than 1M system gates. It is fabricated on thin-epitaxial silicon wafers using 0.22 µ CMOS process with 5 metal layers. The Virtex® family is valuable for space applications because of the reduced cost, high-density and reconfigurability, which can considerably reduce the mission cost. Because it is a VDSM design, it is highly sensitive to radiation effects and its architecture must be studied in order to be protected against upsets.

The Virtex® architecture consists of a flexible and regular matrix composed of an array of configurable logic blocks (CLB) surrounded by programmable input and output blocks (IOB), all interconnected by a large hierarchy of fast and versatile routing resources. The CLB tile is a complex structure composed of Lookup Tables (LUT), flip-flops and routing resources (switch matrix, multiplexors and connection segments), figure 2-8. The CLB provides the functional elements for constructing logic, while the IOB provides the interface between the package pins and the CLB. The logic blocks are interconnected through a general routing matrix (GRM) that comprises an array of routing switches located at the intersections of horizontal and vertical routing channels. The Virtex® matrix also has dedicated memory blocks called select block RAM (BRAM) of 4,096 bits each, clock DLLs for clock-distribution delay compensation and clock domain control, and two 3-state buffers (BUFT) associated with each CLB.

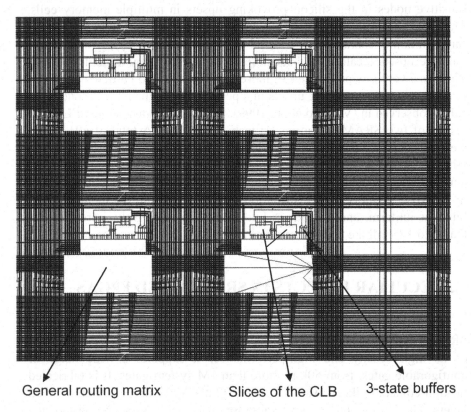

General routing matrix   Slices of the CLB   3-state buffers

Figure 2-8. Example of FPGA architectural model based on the Virtex® family from Xilinx

The CLB tile is composed of CLB slices, where the look-up tables (LUTs) and flip-flops are placed, the single matrix, the hex matrix, the input multiplexers, the output multiplexers, and 3-state buffers, figure 2-9. The single and hex matrixes compose the general routing matrix (GRM).

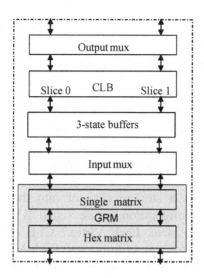

Figure 2-9. The Configurable Logic Block (CLB) Tile of the Virtex® FPGA

Each CLB tile contains two slices. Each slice implements two 4-input Look-Up-Tables (LUTs), two D-Type flip-flops, and some carry logic, figure 2-10. The primary elements in the slices are the F and G LUTs, and the X and Y flip flops. The slices also have internal multiplexers to control the connectivity of internal resources. Finally, there is logic inside each slice to implement fast carries for arithmetic-type logic.

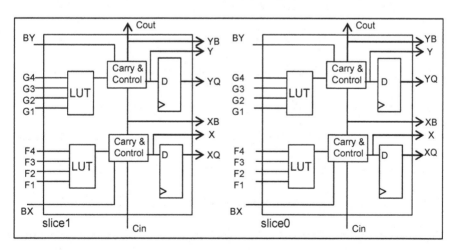

Figure 2-10. Simplified schematic version of CLB Slices 0 and 1 of Virtex® FPGA

Each CLB slice can implement any two of all 4-input logic functions or some functions up to 9 inputs. The function generator is implemented as a lookup table (LUT), figure 2-11. Besides operating as a function generator, each LUT can provide a 16 x 1-bit synchronous RAM. Furthermore, the two LUTs within a slice can be combined to create a 16 x 2-bit or 32 x 1-bit synchronous RAM, a 16 x 1-bit dual-port synchronous RAM, or a 6-bit shift register.

The storage elements in the Virtex® slice can be configured either as edge-triggered D-type flip-flops or as level-sensitive latches. The D inputs can be driven either by the function generators within the slice or directly from slice inputs, by passing the function generators. In addition to Clock and Clock Enable signals, each slice has synchronous set and reset signals (SR and BY). SR forces a storage element into the initialization state specified for it in the configuration, while signal BY forces the storage element into the opposite state. Each Virtex® CLB contains two 3-state buffers (BUFT) that can drive on-chip busses. Each Virtex® BUFT has an independent 3-state control pin and an independent input pin.

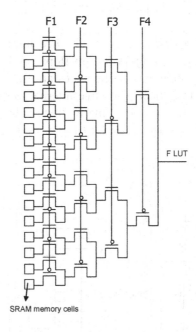

Figure 2-11. 4-input LUT Schematic

The general routing matrix (GRM) uses two kinds of wires: single-length and hex-length wires, figure 2-12. Single wires terminate at an adjacent CLB, while hex wires terminate at CLBs 6 positions over. Single wires should be used to transport data between local CLBs, whereas hex wires should be used to transport data to non-local CLBs. The single and hex wires are each grouped into busses that extend in the four primary directions: north, east, south, and west. The connections to neighboring CLBs are straightforward. A north single wire connects directly to a south single wire in the CLB above it. A hex west wire connects directly to a hex east wire on the 6th CLB over.

The single switch box enables the single wires to be connected together. For example, a single on the north bus can be connected to a single on the west bus or to a single on the south bus. The hex switch box enables the hex wires to be connected together. Using the hex switch box, a signal can traverse a hex wire 6 CLBs up and then 6 CLBs to the west. To summarize, the GRM allows the following connections: single and hex to be connected, single wires to connect to the input multiplexers and single wires to connect to the output multiplexers. The connections are done by Programmable Interconnect Points (PIP) controlled by bits of the bitstream, figure 2-13.

The single wires are numbered from 23 to 0 for each direction. When single wires are connected to other single wires, the PIP is named Single To Single wire. All single to single wire connections are bidirectional, which means that they operate in both directions. In general, all the single wires can be connected to their single wire counterpart (single south to single north and single east to single west), with two more single wires and two hex wires. When single wires are connected to hex wires, the PIP is called Uni Hex To Single wire (connected to a unidirectional hex wire) or Bi Hex To Single (bidirectional hex wire).

The hex wires use multiplexers and buffers for switching, figure 2-14. The hex wires can be unidirectional in, which only drive data into the CLB, or they can be unidirectional out, which can only drive data out of the CLB. In addition, some hex wires can drive data in or out, these are bidirectional. The circuit should drive data on the bidirectional wires in only one direction, not both, since this leads to contention which can damage or destroy the device. The hex wires are numbered from 11 to 0 for each direction. When they are bidirectional, they are numbered 3-0, and when unidirectional they are numbered 11-4.

A peculiarity about the hex wires is that they can only be guided in the extremes: in the first and in the 6th CLB. But they can be read by the GRM in the middle CLB, the third one of the beginning. The two extreme points of any horizontal buffered hex wire connection are named hex_west_# or hex_east_#, and the middle tapped wire connections are hex_horiz_m#, where # ranges from 11 to 0. Vertical hex wires connecting GRMs have the same behavior, hex_north_#, hex_south_# and hex_vert_m#.

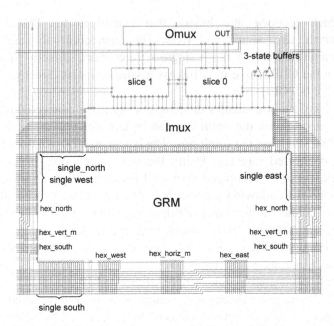

Figure 2-12. Switch matrix connects the Single and Hex Segments

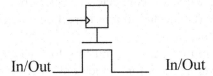

In/Out   In/Out

Figure 2-13. Programmable Interconnect Points (PIPs), where the square represents the
SRAM memory cell that controls the pass transistor

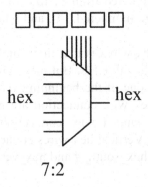

hex   hex

7:2

Figure 2-14. Hex line connections in the routing

The input multiplexers (IMUX) are responsible to connect the input slices to the GRM. There are 13 inputs per slice, which includes the F1-F4, G1-G4, CLK, SR, etc. Each input has a multiplexer associated with it that determines which wire drives the input.

In turn, the output multiplexers (OMUX) are responsible for connecting the output slices to the OUTPUT wires connected to the GRM. There are 8 output multiplexers that connect to the follow outputs (Out0-Out7) per CLB, figure 2-15. Each output multiplexer can select various slice outputs and drive those signals onto the general routing. A slice output can be fanned out to multiple locations by driving multiple output multiplexers.

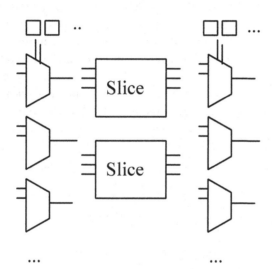

Figure 2-15. Input and Output multiplexors in the routing, where the squares represent the SRAM memory cells

Table 2.1 shows the total number of customization bits for a CLB tile in the Virtex® FPGAs. Note that 35% of the customization bits correspond to the PIPs that control the single wires connections in the single matrix. Analyzing the hex matrix, the number represents all the customization bits, which control multiplexers and buffers. There are 3 bits to control each multiplexer and 1 additional bit for each hex bidirectional wire to control the buffer. The bits able to control the input and the output multiplexers represent roughly 33% of the customization bits, almost same amount of PIPs. The bits used to set up the logic (LUTs) are the minority, only 14%.

Table 2-1. Number of CLB Tile Customization Bits

Classification	Number of bits (percentage)
Output mux	56 (6,48%)
CLB slices	125 (14,47%)
3-state buffers	14 (1,62%)
Input mux	231 (26,74%)
Single matrix	304 (35,18%)
Hex matrix	112 (12,96%)
Not available	22 (2,55%)

Virtex® family has several large Select block RAM (BRAM) memories. Each embedded memory can be programmed with up to 4,098 bits and a single or dual port mode, figure 2-16. These blocks complement the distributed LUT RAMs that provide shallow RAM structures implemented in CLBs. BRAM memories are organized in columns. All Virtex® components contain two such columns, one along each vertical edge. These columns extend the full height of the chip. Each memory block is four CLBs high, and consequently, a Virtex® component 64 CLBs high contains 16 memory blocks per column, and a total of 32 blocks. In the Virtex®-E there are four BRAM columns in the matrix.

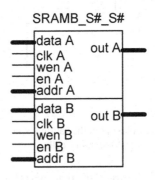

Figure 2-16. Embedded Block RAMs (BRAM)

Virtex® components are programmed by loading a configuration bitstream (collection of configuration bits) into the matrix. The device functionality can be changed at anytime by loading in a new bitstream. The bitstream is divided into frames and it contains all the information to configure the programmable storage elements in the matrix located in the Look-up tables (LUT) and flip-flops, CLBs configuration cells and interconnections and embedded memories. All these bits are potentially sensitive to SEU and consequently they were our investigation targets.

SEU has a peculiar effect in FPGAs when a particle hits the user's combinational logic. In an ASIC, the effect of a particle hitting either the combinational or the sequential logic is transient; the only variation is the time duration of the fault. A fault in the combinational logic is a transient logic pulse in a node that can disappear according to the logic delay and topology. In other words, this means that a transient fault in the combinational logic may or may not be latched by a storage cell. Faults in the sequential logic manifest themselves as bit flips, which will remain in the storage cell until the next load.

On the other hand, in a SRAM-based FPGA, both the user's combinational and sequential logic are implemented by customizable logic memory cells, in other words, SRAM cells, as represented in figure 2-17. When an upset occurs in the combinational logic synthesized in the FPGA, it corresponds to a bit flip in one of the LUTs cells or in the cells that control the routing. An upset in the LUT memory cell modifies the implemented combinational logic, see figure 2-18.

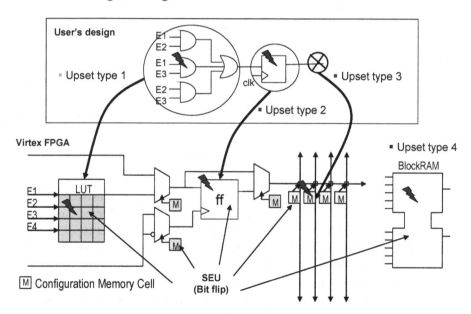

Figure 2-17. The comparison of the effects of a SEU in ASIC and FPGA architecture

It has a permanent effect and it can only be corrected at the next load of the configuration bitstream. The effect of this upset is related to a stuck-at fault (one or zero) in the combinational logic defined by that LUT (figure 2-17, upset type 1). This means that an upset in the combinational logic of a

FPGA will be latched by a storage cell, unless some detection technique is used. An upset in the routing can connect or disconnect a wire in the matrix, see figure 2-19. It has also a permanent effect and its effect can be mapped to an open or a short circuit in the combinational logic implemented by the FPGA (figure 2-17, upset type 3). The fault can also be corrected at the next load of the configuration bitstream.

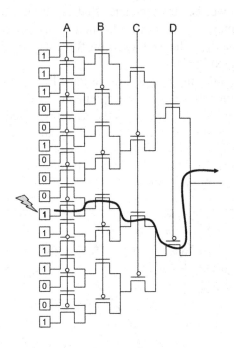

Figure 2-18. Upset in the LUT (logic change)

When an upset occurs in the user sequential logic synthesized in the FPGA, it has a transient effect, because an upset in the flip-flop of the CLB is corrected by the next load of the flip-flop (figure 2-17, upset type 2). An upset in the embedded memory (BRAM) has a permanent effect, and it must be corrected by fault tolerant techniques applied in the architectural or in the high-level description, as the load of the bitstream cannot change the memory state without interrupting the normal operation of the application (figure 2-17, upset type 4). In (Rebaudengo et al., 2002a; Caffrey et al., 2002; Bernardi et al., 2004), the effects of upsets in the FPGA architecture are also discussed. Note that there is also the possibility of having single event transient (SET) in the combinational logic used to build the CLB such as input and output multiplexors used to control part of the routing.

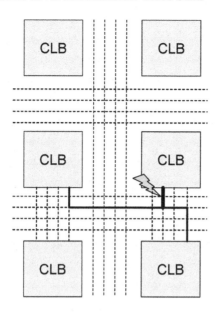

Figure 2-19. Upset in the routing (undesirable connection)

Radiation tests performed in Xilinx FPGAs (Alfke and Padovani, 1998; Katz et al., 1997; Lum and Martin, 1998; Fuller et al., 2000; Sturesson et al., 2001; Fuller et al., 2002) show the effects of SEU in the design application, and prove the necessity of using fault-tolerant techniques for space applications. In (Ohlsson et al., 1998), the effect of neutrons was also analyzed in a SRAM-based FPGA from Xilinx. In that time, the FPGA presented very low susceptibility to neutrons, but the vulnerability is increasing as the technology is reaching smaller transistor size and consequently higher logic density. Experiments with hundreds of latest generation FPGAs operating in tandem on the same board located at high altitude have shown one upset each 2 or 3 months due to neutrons. This number increases with the advance of technology.

A fault-tolerant system designed into SRAM-based FPGAs must be able to cope with the peculiarities mentioned in this section such as transient and permanent effects of a SEU in the combinational logic, short and open circuit in the design connections and bit flips in the flip-flops and memory cells.

Chapter 3

SINGLE EVENT UPSET (SEU) MITIGATION TECHNIQUES
State-of-the-Art

The first SEU mitigation solution that has been used for many years in spacecraft was shielding, which reduces the particle flux to very low levels, but it does not completely eliminate it. This solution was sufficient to avoid errors caused by radiation effects for many years in the past. However, due to the continuous evolution of the fabrication technology process, as explained in last chapter, electronic circuits are becoming more and more sensitive to radiation particles, and the charged particles that once were negligible are now able to cause errors in the electronic design. Consequently, extra techniques must be applied to avoid radiation effects.

Several SEU mitigation techniques have been proposed in the last few years in order to avoid faults in digital circuits, including those implemented in programmable logic. They can be classified as:

- Fabrication process-based techniques, such as:

    • Epitaxial CMOS processes

    • Advanced process such as silicon-on-insulator (SOI).

- Also there are Design-based techniques, such as:

    • Detection techniques:

        ▪ Hardware redundancy

        ▪ Time redundancy

        ▪ EDC (error detection coding)

        ▪ Self-checker techniques

- Mitigation techniques:
    - Triple Modular Redundancy (TMR)
    - Multiple redundancy with voting
    - EDAC (Error detection and correction coding)
    - Hardened memory cell level
- And Recovery Techniques (applied to programmable logic only), such as:
    - Reconfiguration
    - Partial configuration
    - Rerouting design

The fabrication process-based techniques, also called technological techniques, such as epitaxial CMOS process and silicon-on-insulator (IBM, 2000; Colinge, 2001; Musseau and Ferlet-Cavrois, 2001), can reduce to acceptable levels some of the radiation effects, such as Total Ionization Dose (TID) effects and single event latch-up (SEL), however, they do not completely eliminate upset effects, such as single event upsets (SEUs) and single transient effect (SET). The fabrication process-based solution is expensive and consequently very few designs have adopted this technique, especially for low volume production. In (Irom et al., 2002), SEU effects from heavy ions and protons are measured for Motorola and IBM silicon-on-insulator microprocessors, and compared with results from similar devices with bulk substrates. Results show that the threshold LET values of the SOI processors are nearly the same as those of bulk/epi processors from the same manufactures, indicating that little improvement in SEU sensitivity has resulted from the move to SOI technology. Although the threshold LET did not change, the cross section of the SOI processors were about an order of magnitude lower than the bulk/epi counterparts, leading to a lower upset rate in the space environment. These results show that only modest improvements in SEU sensitivity can be expected as mainstream integrated circuits move to SOI technology, and consequently design-based techniques must be applied to mitigate SEU.

The design-based techniques, also called architectural techniques, are highly accepted because they can be applied to many different levels of the design without any changes in the fabrication process technology. They can be planned to just detect the presence of an upset in the system or they can be more complex in order to detect and correct the system error in the presence of an upset. All design-based techniques are composed of some kind of redundancy, which can be provided by extra components (hardware redundancy) or by an extra execution time or by different instants of data sampling (time redundancy). Very often, techniques implement a combination of both.

Hardware redundancy is basically based on logic redundancy, which is characterized as extra components or extra paths that allow the design to continue operation even when some parts fail. Error detection and correction codes (EDAC) can also been seen as a hardware redundancy because it generates redundant bits to be able to detect and correct upsets. EDAC codes can be also used in time redundancy techniques, as it is going to be discussed in next sections. By concerning only the SEU problem in storage devices, hardened memory cells is another example of hardware redundancy that can be applied to avoid error in memories.

Recently, new techniques based on recovery have been proposed particularly for programmable logic components, such as SRAM-based FPGAs. The idea is to recover the original programmed information after an upset. Examples of this technique are reconfiguration (scrubbing), partial reconfiguration and rerouting design. They are able to clean out an upset in the programmable matrix in a very short period of time. This type of technique is usually used to avoid the accumulation of upsets.

Finding the most appropriate SEU mitigation solution has become a challenge in order to combine fast turnaround time, low cost, high performance and high reliability. An efficient set of SEU mitigation techniques should cope with transient faults (SET) occurring in the combinational logic and SEUs in the storage cells. In this way, transient faults in the combinational logic will never be stored in the storage cells, and bit flips in the storage cells will never occur or will be immediately corrected. Each technique has some advantages and drawbacks, and there is always a compromise between area, performance, power dissipation and fault tolerance efficiency.

This chapter presents an overview of the design-based techniques on digital circuits, and subsequently it shows the state o the art of SEU mitigation techniques for ASICs and FPGAs, including solutions using the recovery method.

3.1 DESIGN-BASED TECHNIQUES

Time and hardware redundancy techniques are largely used in ASICs (Nicolaidis, 1999; Dupont et al., 2002; Benz et al., 2002). The techniques range from simple upset detection to upset voting and correction. There is a wide choice of techniques according to the user's application requirements.

Sometimes it is just necessary to warn the presence of an upset with an interruption in the system functionality, while sometimes it is required to completely avoid interruptions, assuring full reliability. There is a set of techniques that can present reliability in between these two extremes, each one producing more or less overhead according to its robustness in the presence of a fault.

3.1.1 Detection Techniques

Techniques based on time redundancy are usually used to detect a transient effect (SET) in the combinational logic, while hardware redundancy can help to identify an SEU in the sequential logic. Examples of the use of time and hardware redundancy for SET detection have been presented in the (Nicolaidis, 1999; Anghel et al., 2000; Dupont et al., 2002). In the case of time redundancy, the goal is to take advantage of the characteristics of the transient pulse generated by the particle strike to compare the output signals at two different moments. The output of the combinational logic is latched at two different times, where the clock edge of the second latch is shifted by time d. A comparator indicates a transient pulse occurrence (error detection). The scheme is illustrated in figure 3-1.

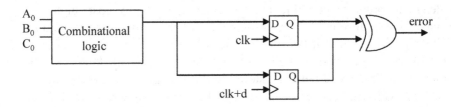

Figure 3-1. Time redundancy duplication scheme to detect SET in combinational logic

In the case of hardware redundancy, the duplication with comparison (DWC) scheme can be used for both combinational and sequential logic to SET and SEU detection, respectively. Figures 3-2 and 3-3 show the schemes for transient effect detection. Note that for both techniques, time and hardware redundancy, it is important to take into account the duration of the SET.

Another example of upset detection for sequential logic is error-detecting codes such as parity bits. In this case, the parity bit of a group of analyzed bits is calculated and it is continuously compared to a new parity bit calculation. If a SEU has occurred, it is possible to detect it. This solution is largely used nowadays in memories. However, for high-reliability applications, sometimes it is not enough only to detect the presence of a fault, rather one has to ensure the correct operation of the system in the presence of that fault. For this reason, it is very important to investigate in detail the SEU mitigation solutions.

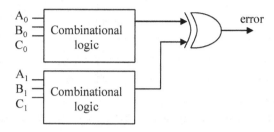

Figure 3-2. Hardware redundancy duplication scheme to detect SET in combinational logic

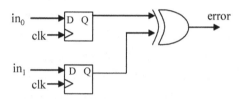

Figure 3-3. Hardware redundancy duplication scheme to detect SEU in sequential logic

3.1.2 Mitigation Techniques

3.1.2.1 Full Time and Hardware Redundancy

The use of full time redundancy in the combinational logic permits voting the correct output value in the presence of a SET. The name full redundancy comes from the complete n-modular redundancy, where n is equal to three: triple modular redundancy. In this case, the output of the combinational logic is latched at three different moments, where the clock edge of the second latch is shifted by the time delay d and the clock of the third latch is shifted by the time delay $2.d$. A voter chooses the correct value. The full time redundancy scheme is illustrated in figure 3-4 and the majority voter scheme is presented in figure 3-5. The area overhead comes from the extra sample latches and the performance penalty is given by clk+$2.d$+t_p, where d depends on the duration of the transient current pulse and t_p is the delay from the majority voter. The total delay is measured by the error pulse width multiplied by 2, which is approximately equal to $2 \times (Q_{COL} / I_D)$.

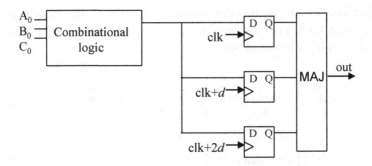

Figure 3-4. Full time redundancy scheme to correct SET in combinational logic

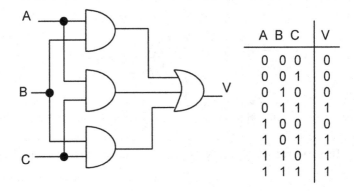

A	B	C	V
0	0	0	0
0	0	1	0
0	1	0	0
0	1	1	1
1	0	0	0
1	0	1	1
1	1	0	1
1	1	1	1

Figure 3-5. Majority Voter Schematic and the Truth Table

In the case of the full hardware redundancy, the well-known Triple Modular Redundancy (TMR) approach, the logic is triplicated and voters are placed at the output to identify the correct value. The first possibility that was largely used in space applications is the triplication of the entire device, figure 3-6. This approach uses a voter as a fourth component in the board. It needs extra connections and it presents area overhead. If an error occurs in one of the three devices, the voter will choose the correct value. It protects both combinational and sequential logic against upsets. However, if an upset occurs in the voter, the TMR scheme is ineffective and a wrong value will be present in the output. Another problem of this approach is the accumulation of upsets, hence an extra mechanism is necessary to correct the upset in each device before the next SEU happens.

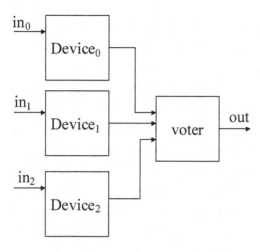

Figure 3-6. TMR implemented in the entire device

A more efficient implementation of the TMR is applied focused in the sensitive logic, for example the memory cells to protect against SEU, see figure 3-7.

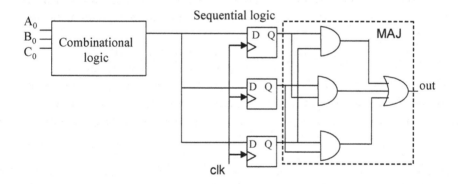

Figure 3-7. TMR Memory Cell with Single Voter

However, this solution does not avoid the accumulation of upsets in the sequential logic and the voter is vulnerable to upsets. In order to restore the corrected value, a solution using three voters with a feedback was proposed (Carmichael, 2001; Katz et al., 2001), figure 3-8. The upsets in the latches are corrected by extra logic in order to avoid accumulation. The load frequency (refreshing) can be set by the multiplexor control signal.

Sequential logic

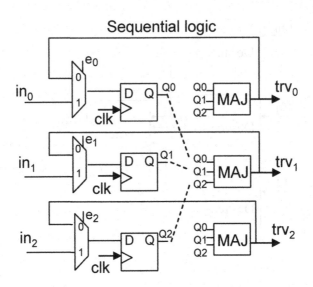

Figure 3-8. TMR memory cell with three voters and refreshing (version I)

The combinational logic must also be protected to avoid SET. There are many possibilities. One is to use time redundancy in the logic as shown in figure 3-9. Another possibility is to triplicate the combinational logic as well, as represented in figure 3-10.

Although the last proposed implementation of the TMR (figure 3-10) presents a larger area overhead compared to time redundancy, since it triplicates all the combinational and sequential logic, it protects the logic against SET and SEU, and avoids accumulation of upsets. In addition, it does not have major performance penalties, just the voter propagation time, and it does not need different clock phases.

Another method to mitigate SET in combinational logic is based on duplication and a code word state preserving (CWSP) (Anghel et al., 2000), as illustrated in figure 3-11. This method does not need voters or comparators. The duplication can be replaced by time redundancy as well, which reduces the area overhead significantly, figure 3-12. The main contribution of this method is the CWSP stage, which replaces the last gates of the circuit by a particular gate topology, which is able to pass the correct value in the combinational logic in the presence of a SET, figure 3-13. Additional techniques to cope with SET are presented in (Alexandrescu et al., 2002).

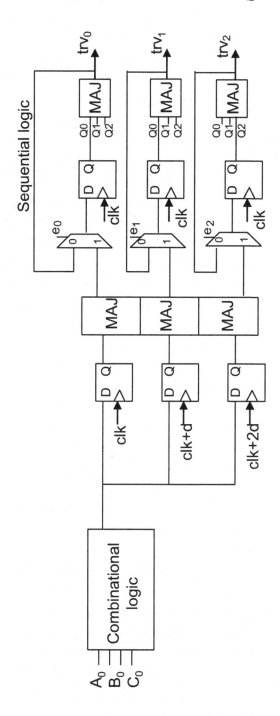

Figure 3-9. Full time redundancy scheme for combinational logic combined to full hardware redundancy in the sequential logic

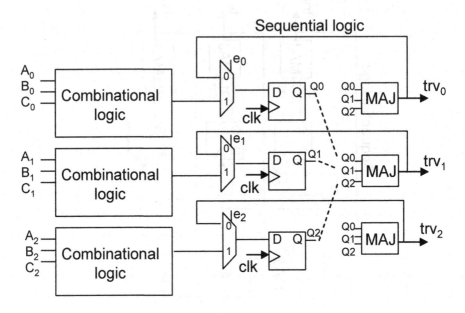

Figure 3-10. Full hardware redundancy scheme for combinational and sequential logic

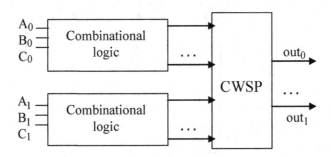

Figure 3-11. Duplication to mitigate SET in combinational logic (Anghel; Alexandrescu; Nicolaidis, 2000)

Some application systems concern about multiple upsets. However the problem of multiple upsets must be carefully analyzed. Solutions are not trivial. For n-modular redundancy, where n is usually an odd integer, solutions with n larger than 3 does not always present gains in reliability compared to the TMR because the result depends on the failure rate: λ. In Shooman 2002, it is shown that n-modular redundancy is only superior to single modules in high-reliability regions.

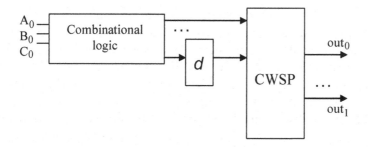

Figure 3-12. Time redundancy to mitigate SET in combinational logic (Anghel; Alexandrescu; Nicolaidis, 2000)

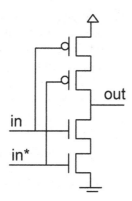

Figure 3-13. Example of INVERTER logic with the code word state preserving (CWSP) in the duplication and time redundancy to mitigate SET in combinational logic

3.1.2.2 Error Correction and Detection Codes

Error detection and correction coding (EDAC) technique (Peterson, 1980) is also used to mitigate SEU in integrated circuits. It is usually used in memory. There are many codes to be used to protect the systems against single and multiple SEUs. An example of EDAC is the hamming code (Houghton, 1997) in its simplest version. It is an error-detecting and error-correcting binary code that can detect all single- and double-bit errors and correct all single-bit errors (SEC-DED). This coding method is recommended for systems with low probabilities of multiple errors in a single data structure (e.g., only a single bit error in a byte of data). The code satisfies the relation $2k \geq m+k+1$, where $m+k$ is the total number of bits in the coded word, m is the number of information bits in the original word, and k is the number of check bits in the coded word. Following this equation the hamming code can correct all single-bit errors on n-bit words and detect double-bit errors when an overall parity check bit is used.

The hamming code implementation is composed of a combinational block responsible for encoding the data (encoder block), inclusion of extra bits in the word that indicate the parity (extra latches or flip-flops) and another combinational block responsible for decoding the data (decoder block). The encoder block calculates the parity bit and it can be implemented by a set of 2-input XOR gates. The decoder block is more complex than the encoder block, because it needs not only to detect the fault, but it must also correct it. It is basically composed of the same logic used to compose the parity bits plus a decoder that will indicate the bit address that contains the upset. The decoder block can also be composed of a set of 2-input XOR gates and some AND and INVERTER gates.

The encoder block calculates the check bits that are placed in the coded word at positions 1, 2, 4, ..., 2(k-1). For example, for 8-bit data, 4 check bits (p1, p2, p3, p4) are necessary, so that the hamming code is able to detect and correct a single-bit error (SEC-SED). Figure 3-14 demonstrates a 12-bit coded word (m = 8 and k = 4) with the check bits p1, p2, p3 and p4 located at positions 1, 2, 4 and 8 respectively. The check bits are able to inform the position of the error. The encoder block can be implemented by a set of 2-input XOR gates. For an 8-bit data, 14 2-input XOR gates are necessary in order to generate the 4 parity bits. The check bit p1 creates even parity for the bit group {1, 3, 5, 7, 9, 11}. The check bit p2 creates even parity for the bit group {2, 3, 6, 7, 10, 11}. Similarly, p3 creates an even parity for the bit group {4, 5, 6, 7, 12}. Finally, the check bit p4 creates even parity for the bit group {8, 9, 10, 11, 12}.

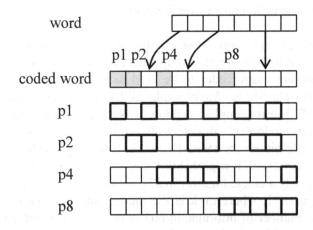

Figure 3-14. Hamming code check bits generation for a 8-bit word, 12-bit coded word

Hamming code can protect structures such as registers, register files and memories. Each protected register must have its input connected to the encoder block, and its output connected to the decoder block. Note that only one register may be used at a clock cycle. The main advantage of the set of registers structure is that only one encoder block and one decoder block are multiplexed for a set of registers.

Hamming code increases area by requiring additional storage cells (check bits), plus the encoder and the decoder blocks. For an n bit word, there are approximately log_2n more storage cells. However, the encoder and decoder blocks may add a more significant area increase, thanks for the extra XOR gates. Regarding performance, the delay of the encoder and decoder block is added in the critical path. The delay gets more critical when the number of bits in the coded word increases. The number of XOR gates in serial is directly proportional to the number of bits in the coded word.

Table 3.1 shows a comparison between hamming code and the full hardware redundancy (TMR) to mitigate SEU in sequential circuits. Results published in (Hentschke et al., 2002) show that TMR is more efficient in terms of area and performance to protect registers and small memory structures, while hamming code is more appropriate to protect large register files and memories.

Table 3-1. Hamming Code and TMR Comparison Summary

	Hamming Code (SEC-DED)	TMR
Area	It depends on the number of bits to be protected. It has a small overhead of storage cells (parity cells) It needs additional combinational logic to implement the encoder and the decoder blocks in the case of short coded words.	It needs 3 times more storage cells. It needs small extra logic for the voters. The number of voters is proportional to the number of storage cells.
Performance	The encoder and decoder blocks, which are located in the critical path, can affect the performance. The delay increases proportionally to the number of bits to be coded because of the number of XOR	The performance is not strongly affected because the only source of delay is the voter that is basically constant with the number of bits to be protected.

	Hamming Code (SEC-DED)	TMR
	gates in serial in the encoder and decoder blocks.	
Error-correcting code	It corrects one single upset per word. But it does not correct the upset in the stored word. Upsets will accumulate if there is no extra logic to correct them.	It corrects up to n upsets per n-bit word if each upset is located in a distinct bit. It votes the correct value but it does not correct it. Upsets will accumulate if there is no extra logic to correct them.

The problem of hamming code is that it can not correct double bit upsets, which can be very important for very deep sub-micron technologies, especially in memories because of the high density of the cells (Reed et al., 1997). Other codes must be investigated to be able to cope with multiple bit upsets. Reed-Solomon (Houghton, 1997) is an error-correcting coding system that was devised to address the issue of correcting multiple errors. It has a wide range of applications in digital communications and storage. Reed-Solomon codes are used to correct errors in many systems including: storage devices, wireless or mobile communications, high-speed modems and others. Reed-Solomon (RS) encoding and decoding is commonly carried out in software, and for this reason the RS implementations normally found in the literature do not take into account area and performance effects for hardware implementation. However, the RS code hardware implementation as presented in (Neuberger et al., 2003) is an efficient solution to protect memories against multiple SEUs.

A Reed-Solomon code is specified as RS(n, k) with s-bit symbols, where n is the total number of symbols per coded word and k is the number of symbols per information data. The number of parity symbols is equal to n − k, where n is 2 raised to the power of s minus one ($2s - 1$). A Reed-Solomon decoder can correct up to t number of bytes, where $2t = n - k$, figure 3-15.

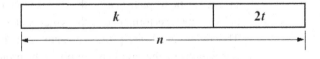

Figure 3-15. Reed-Solomon coded word

Mathematically, Reed-Solomon codes are based on the arithmetic of finite fields. In the case of applying RS code in memories, the data word is divided in symbols, and each data word is a different RS coded word. For example, in an n-rows memory, the data word uses the entire row, and each data word is divided in m symbols according to the symbol size and to the memory data size. Multiple upsets may occur in any portion of the matrix, but they are more likely to occur as double bit flips that are in the same symbol (upset type a), in vertical adjacent symbols, (upset type b), or in horizontal adjacent symbols, (upset type c), as shown in figure 3-16.

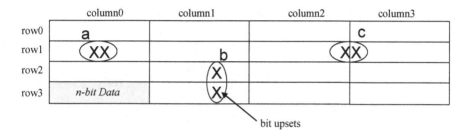

Figure 3-16. Examples of double bit flips in a memory where each row is protected by RS code

The RS code can easily correct upsets of type a, because it is the essential property of this code: multiple error correction in a same symbol. The second type of double upsets that can occur, upset type b illustrated in the figure 3-16, will also be corrected because each row is a different RS code, so this is equivalent to two single errors in distinct rows. But the third type of upsets, upset type c, illustrated in figure 3-16, will not be corrected, because it is equivalent to errors in two different symbols of the same coded word, and the implemented RS is not capable to correct this type of error. In the next chapter, a solution for this problem is proposed and some results of protecting a memory component with this new solution based on RS code are discussed.

3.1.2.3 Hardened Memory Cells

Another example of SEU mitigation technique is memory cells composed of extra devices, which can be resistors or transistors, able to recover the stored value if an upset strikes one of the drains of a transistor in "off" state. These cells are called hardened memory cells, and they can avoid the occurrence of a SEU by design, according to the flux and to the charge of the particle.

In order to better understand how these hardened memory cells work, let us start with the analysis of a standard memory cell composed of 6 transistors (figure 3-17). When a memory cell holds a value, it has two transistors in "on" state and two transistors in "off" state; consequently there are always two SEU sensitive nodes in the cell. When a particle strikes one of these nodes, the energy transferred by the particle can provoke a transistor to switch "on". This event will flip the value stored in the memory. If a resistor is inserted between the output of one of the inverters and the input of the other one, the signal can be delayed for such a time to avoid the bit flip.

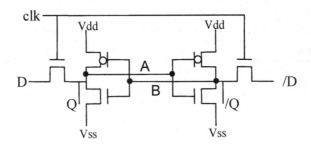

Figure 3-17. Standard Memory Cell

The SEU tolerant memory cell protected by resistors (Weaver et al., 1987) was the first proposed solution in this matter, figure 3-18. The decoupling resistor slows the regenerative feedback response of the cell, so the cell can discriminate between an upset caused by a voltage transient pulse and a real write signal. It provides a high silicon density, for example, the gate resistor can be built using two levels of polysilicon. The main drawbacks are temperature sensitivity, performance vulnerability in low temperatures, and an extra mask in the fabrication process for the gate resistor. However, a transistor controlled by the bulk can also implement the resistor avoiding the extra mask in the fabrication process. In this case, the gate resistor layout has a small impact in the circuit density.

Memory cells can also be protected by an appropriate feedback devoted to restore the data when it is corrupted by an ion hit. The main problems are the placement of the extra transistors in the feedback in order to restore the upset and the influence of the new sensitive nodes. Examples of this method are IBM hardened memory cells (Rockett, 1988) in figure 3-19, HIT cells (Bessot and Velazco, 1993; Velazco et al., 1994; Calin et al., 1996) in figure 3-20 and Canaris memory cells (Wiseman et al., 1993; Canaris and Whitaker, 1995) in figure 3-21 and 3-22. The main advantages of this method are temperature, voltage supply and technology process independence, and good

SEU immunity. The main drawback is silicon area overhead that is due to the extra transistors and their extra size.

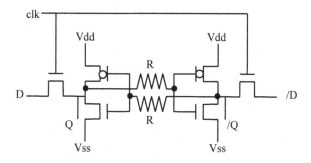

Figure 3-18. Resistor Hardened Memory Cell

The IBM cell has 6 extra transistors, figure 3-19, PA and PB are called data state control transistors, PC and PD are pass-transistors and PE and PF are cross-coupled transistors. The sensitive nodes are A, B, and C. When a particle hits the node A, it instantly goes low, and momentarily the cell is unstable with both nodes A and B at a relative low potential. Transistor PD momentarily turns on but node D cannot charge low enough to turn PB fully ON, since transistor PF remains ON. However the presence of the fully ON PA transistor, reinforcing the pre-hit relatively positive data state at node A, restores node A without logic upset.

Considering now a particle hit occurring at node B, when the hit occurs node B goes high turning transistor PC OFF, momentarily isolating node C at its relative low potential. With the gate of transistors P1 and N1 connected to node B, the resulting data feedback response causes node A to attempt to go low. However, with the transistor PA ON reinforcing the preexisting high state in node A, node A maintains its high state data. Therefore node B eventually returns to its pre-hit low potential after the momentary disturbed condition, the transistor N2 once again pulls down node B. Thus node B recovers the logic upset.

Finally, if a particle hits node C, transistors PA and PF momentarily turn off. With respect to data information stored in the data cell, no harm is done, and node C is eventually recharged low through the ON PC transistor. Node C recovers and there is no threat posed to the stored data.

The HIT cell has also 6 extra transistors placed in a feedback around the main storage cell, figure 3-20. In the normal operation, if the read/write signal is low (inactive) transistors MP1, MP4, MN2, MN6 and MP5 are ON, the other transistors being OFF. Then, it is easy to show that the logical states of nodes Q and /Q are conserved. Furthermore, as there are no direct paths from Vdd to Vss, the stability of the HIT cell memorization function is guaranteed.

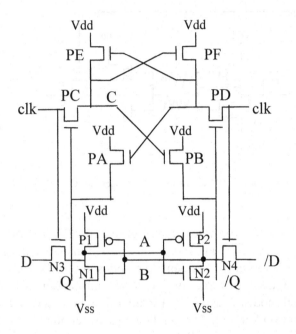

Figure 3-19. IBM Hardened Memory Cell

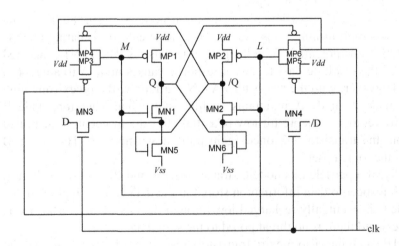

Figure 3-20. HIT Hardened Memory Cell

Read operation is performed by pre-charging to VDD data lines D and /D. As the read/write signal goes high, Q will remain at 1 because it is directly connected to the data line D through transistors MN1 and MN3. Node /Q will remain at 0 because MN4 and MN6 are both ON discharging data line /D.

The HIT cell has 3 sensitive nodes that are Q, /Q and M. If a particle strikes the drain of transistor MN1, node Q will go low. Transistors MN6 and MP6 will turn OFF and ON respectively. Then, node /Q is not biased, but conserves its low state by capacitive effect. Transistors MP6 and MP5 are both ON but, as the width of MP5 is chosen larger than the width of MP6, node L will remain at 1. As transistor MP1 is still ON, node Q will be restored to 1, recovering the upset.

If a particle strikes the drain of the transistor MP2, node /Q will go to 1, turning transistors MN5 and MP4 respectively ON and OFF. Node M goes to high impedance, conserving its initial 0 state. As transistors MN2 and MN6 are still ON, node /Q is restored to its initial 0 state.

If the drain of transistor MP3 is hit by a particle, node M will go high, turning ON and OFF transistors MN1 and MP1 respectively. As transistors MN5 and MP5 are OFF, nodes Q and L become at high impedance conserving their states. As /Q is still low, transistor MP4 will remain ON restoring the state of node M that goes to 0.

Multiple SEUs can occur in the memory cell. HIT cells cannot manage with this problem. For example, if a particle strikes on M, this leads to turn OFF transistors MP1 and MP5, and if another particle strikes on Q, it turns transistors MN6 and MP6 OFF and ON respectively. Then node L is pulled down turning ON and OFF transistors MP2 and MN2 respectively and /Q goes high, turning ON and OFF transistors MN5 and MP5 respectively. Node M is asserted to 1 and node Q is asserted to 0. The contents of the memory cell are then corrupted. In a similar way it can be shown that simultaneous particle strikes on these nodes /Q and M lead to the corruption of the data stored in the memory cell. SEU testing presented in (Velazco et al., 1994) shows that the hardened HIT cell design is less sensitive at least by a factor of 10 than unhardened cell design.

The Canaris approach consists of a memory cell built with AND-NOR and OR-NAND gates that are SEU immune. Each logic gate has two outputs, one for the N-channel transistor and other for the P-channel transistors. Transistor M1 is sized to be weak compared to the p-channel array and transistor M2 is sized to be weak compared to the n-channel array in such way that it can be restored to the original value in the output if a particle hits the sensitive nodes.

The interesting aspect of this solution is that it can be applied to even the combinational and sequential logic when memory cells are implemented using the SEU immune combinational gates. Using this approach, all the combinational part of the circuit can be grouped in complex logic functions where each one of these functions has two extras transistors dividing their outputs. For large complex logic gates, two extra transistors may not represent a high addition of area. However, due to the duplication of outputs the number of internal connections can increase according to the implementation architecture. The main drawback of Canaris hardened memory cell is the long recovery time after upset.

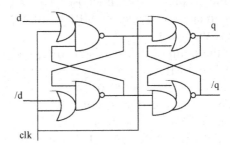

Figure 3-21. Canaris SEU hardened memory cell

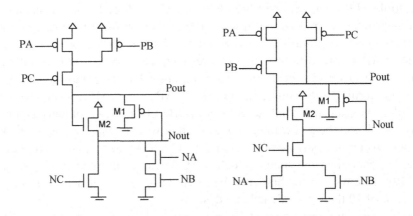

Figure 3-22. Canaris SEU hardened memory cell detailed implementation

Another mitigation principle is to store the data in two different locations in the cell in such a way that the corrupted part can be restored. Examples of this technique are DICE cells (Canaris and Whitaker, 1995) in figure 3-23 and NASA cells (Whitaker et al., 1991; Liu and Whitaker, 1992) in figure 3-24 and 3-25 respectively. The main advantages of this method are also temperature, voltage supply and technology process independence, good SEU immunity and high performance (read/write time).

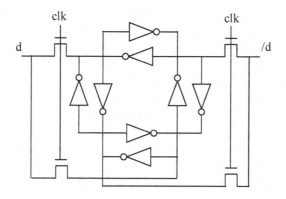

(a) Dice cell represented by inverters

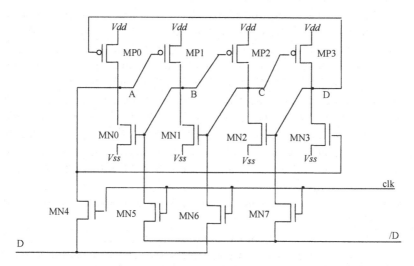

(b) Dice cell represented in detail
Figure 3-23. DICE Hardened Memory Cell

DICE cell consists of a symmetric structure of four CMOS inverters, where each inverter has the n-channel transistor and the p-channel transistor separately controlled by two adjacent nodes storing the same state, figure 3-23 (b). The 4 nodes of the DICE cell form a pair of latches in two alternate ways, depending on the stored logic value. One of the adjacent nodes controls the conduction state of the transistor connecting the current node to a power supply line, and the other node blocks the complementary transistor of the inverter, isolating it from the opposite supply line.

In Figure 3-23(b), the adjacent node pairs A-B and C-D have active cross-feedback connections and form two-transistor, state-dependent latch structures. The other adjacent node pairs, B-C and D-A, have inactive feedback connections (off transistors) which isolate the two latching pairs. Hence, two non-adjacent nodes are logically isolated and must be both reverted in order to upset the cell. If a charged particle hits a sensitive node, it flips the state logic and switches off the active feedback transistor controlling the adjacent latching node. The second node of the latching structure conserves its state by capacitive effect.

The inactive feedback transistor to the adjacent isolated node is switched on, and generates a logic conflict, which is propagated to the second latching node. The active feedback connections from the two unaffected nodes restore the initial state at the upset node and subsequently remove the state conflict of the second perturbed node. A write operation in DICE cell is required to store the same logic state at two non-adjacent cell nodes in order to revert the logic state of the cell.

The NASA cells also store the information in two different places. This provides redundancy and maintains a source of uncorrupted data after a SEU. The recovery path is based on the use of weak and strong transistors. The weak transistor size is approximately 1/3 of the normal transistor size. The size of the weak feedback transistors is responsible for the recovery time. The DICE latch is, in principle, SEU immune in that two nodes must be simultaneously driven to change the state of the latch. A single cosmic ray can, however, simultaneously strike two critical nodes if it passes through the chip at an extremely small angle of incidence. The probability of this occurring depends on the solid angle subtended by drain diffusions and the integral fluence of cosmic rays with an LET (linear energy transfer) value greater than some threshold that depends on the circuit response and collection volume.

Another SEU hardened memory solution is presented in (Mavis and Eaton, 2000), figure 3-26. The hardened memory cell contains nine level-sensitive latches (U1 through U9), one majority gate (U10), and three inverters (U11 through U13). Each level-sensitive latch is transparent (sample mode) when its clock input is high and is blocking (hold mode) when its clock input is low. When in sample mode, data appearing at the input D also appears at the output Q. When in hold mode, the data stored within the latch appears at the output Q and any data changes at the input D are blocked. Two level sensitive latches in tandem and clocked by complementary clock signals (such as U1 followed by U2) form an edge triggered D flip-flop. With the clock inversions, the D-Flip-Flops formed by (U1,U2), (U3,U4), and (U5,U6) are triggered on the falling edges of the clocks CLKA, CLKB, and CLKC, respectively.

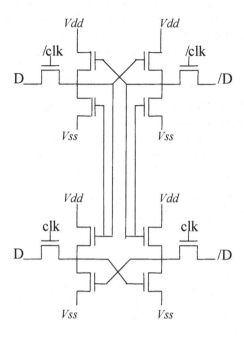

Figure 3-24. NASA I Hardened Memory Cell

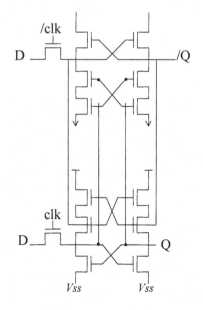

Figure 3-25. NASA II Hardened Memory Cell

Each of these four clocks operates at a 25% duty factor and each one is delayed to the master clock. CLKA is high during the first half of cycle one of the master clock. CLKB is high during the second half of cycle one of the master clock. CLKC and CLKD are high during the first and second halves, respectively, of cycle two of the master clock. Thus a full cycle of the A, B, C, and D clocks occupies two cycles of the master clock. These clocks are actually quite easy to generate with simple circuitry presented in a later section. Controlling the fidelity of the four clocks is not a problem since the level sensitive latch will operate correctly even in the presence of skew or overlaps.

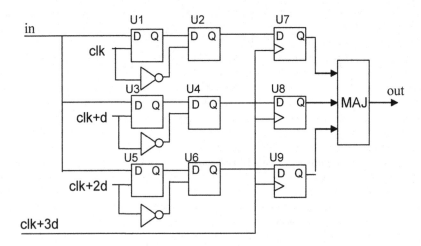

Figure 3-26. Temporal Sampling Latch with Sample and Release Stages

The upset immunity of the circuit in figure 3-26 is a consequence of two distinct parallelisms: (1) a spatial parallelism resulting from the three parallel circuit branches and (2) a temporal parallelism resulting from the unique clocking scheme. In addition, when implemented using DICE-based latches, the temporal latch can achieve immunity to multiple node cosmic ray strikes and, unlike any other SEU mitigation approach, it is immune to a second and third-order effect.

Analyzing the SEU hardened robustness to MBU, the temporal latch, in its simplest form, is clearly immune to upset from any single cosmic ray striking a single circuit node (a first-order effect). This is also true for TMR-based latches and for DICE-based latches. Multiple node strikes (a second-order effect), although having much lower probabilities of occurrence, will surely cause upsets when such latches are fielded in an actual space environment.

Table 3-2 presents a general comparison between the techniques presented in this section: hardened memory cells, hamming code and TMR.

Table 3-2. SEU mitigation techniques summary

SEU Mitigation Technique	SEU Tolerant Memory cells	Hamming Code	TMR
Area	Usually it doubles the area of each memory cell. It is strongly layout and transistor size dependent.	It depends on the number of bits to be protected. It has extra sequential and combinational logic.	It presents a little more than 3 times the area overhead because of the voter.
Performance	The performance is not affected if the extra transistors or resistors (path delay) work only when the cell is on hold.	The encoder and decoder blocks can affect the performance.	The performance is not strongly affected. The only source of delay is the voter.
Error correction	It avoids the error by a delay in the memory loop (redundancy/recovery).	Normally it corrects one single upset per word, but in order to refresh the stored value an extra path is necessary (scrubbing rate).	It does not correct the upsets. The upsets will accumulate if there is no extra logic for the refreshing.
Multiple Upset	Robust to 3rd order of multiple upsets as each cell protects itself.	Not efficient for multiple upsets in the same coded word. But efficient for multiple upsets in different parts of the circuit.	It can be robust for multiple upsets in different parts of the circuit but not in the same TMR signal.
Technology	It can use some extra area because of the asymmetry of the transistors and large resistance in polysilicon.	Completely compatible with CMOS technology.	

3.2 EXAMPLES OF SEU MITIGATION TECHNIQUES IN ASICS

Many commercial microprocessors from Intel, IBM, Motorola and Sun are available in the market in a radiation tolerant version. These hardened microprocessors were designed by space project companies and research laboratories. The fault tolerance concern has started many years ago (Sexton, 1991; Hass et al., 1989). Each product offers different levels of radiation Immunity for distinct space and military applications. The techniques used to protect the microprocessors are usually based on the process technology or package shielding, TMR, SEU hardened memory cells, EDAC (hamming code) or a combination of them.

In (Lima et al., 2000a, 2000b), a radiation fault-tolerant version of the 8051-like micro-controller (Intel, 1994) is proposed. This work was started based on the testing techniques and the studies about EDAC codes published in (Cota et al., 1999). The VHDL (Skahill, 1996) description of this micro-controller was designed at UFRGS (Carro et al., 1996; Silva et al., 1997) and it was re-used to insert SEU radiation fault-tolerant structures. The original code is entirely compatible with the INTEL 8051 microprocessor in terms of instruction timing. The microprocessor description is divided into six main blocks. These units are finite state machine, control unit, instruction unit, datapath and RAM and ROM memories. Single error correction hamming code (SEC) was applied in all registers and internal memory as represented in figure 3-27.

This technique was innovative because it uses EDAC not only in the memory but also in all registers and single memory cells. The memory has a refreshing mechanism, called scrubbing, to avoid accumulation of upsets. A detailed scheme of the hamming code implementation is presented in figure 3-28 and 2-29.

A board implementation has been done with the robust 8051, figure 3-30. The hardened description was prototyped into three programmable logic devices customizable by EEPROM technology from Altera, family MAX 9000, one EPM9560 with 208 pins and two EPM9400 with 84 pins (Altera, 2001). The SEU hardened 8051 daughter board has been tested in the THESIC tester environment (Velazco et al., 2000) under radiation conditions in Louvain-la-Neuve (Belgium) using the Cyclone radiation facility. Cyclone is a cyclotron offering the possibility of accelerating various heavy ion species. Two versions of the 8051 were implemented in the board: 1) the standard 8051 version without protection and 2) the 8051 with the internal memory protected by hamming code.

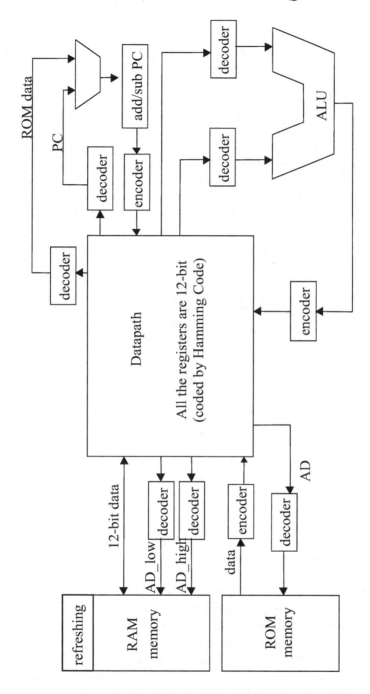

Figure 3-27. General scheme of the SEU hardened 8051

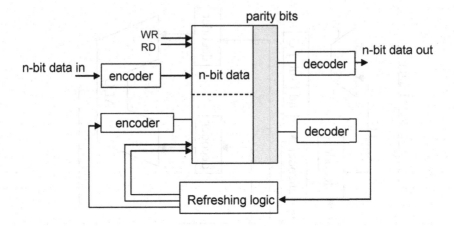

Figure 3-28. Scheme of the hamming code implemented in the memory of a micro-controller

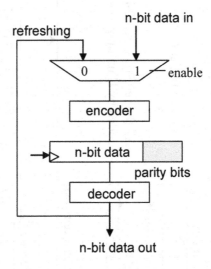

Figure 3-29. Scheme of the hamming code implemented in registers of a micro-controller

The application test of the standard 8051 without protection shows that many upsets have occurred in each analyzed period of time. Figure 3-31 shows the number of errors of each period analysis, for a flux of 700 particles per second. The same application test of the 8051 with the internal memory protected by hamming code showed that "NO ERROR" has occurred for the radiation energy mentioned before. The result proves the efficiency of the hamming code method in SEU protection.

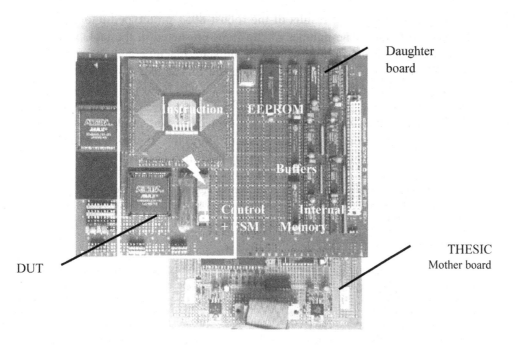

Figure 3-30. SEU Hardened 8051 daughter board and THESIC mother board

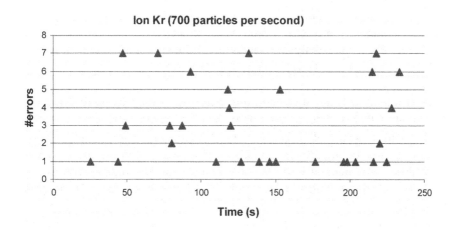

Figure 3-31. Radiation Test result I of the "Not protected" 8051 in the matrix multiplication test

Table 3-3 shows the results of the robust 8051 prototyped into the PLD MAX9000 family. The number of flip-flops presented in the table refers to the internal registers of control unit, finite state machine and datapath. The internal memory is implemented outside the PLDs, like it is shown in the board photo. The full-protected versions of the 8051 did not fit in the PLDs family due to the reduced number of CLBs in the MAX9000 family. Consequently, only the datapath partial protection was implemented in the board. In the partial protected datapath only the accumulator and the program counter registers are protected by hamming code.

Table 3-3. Results of robust 8051-like Micro-Controller implemented in PLDs

1 Version	Control Unit	State Machine	Internal Memory	Datapath	#flip-flops	#CLBs
A	Not protected	Not protected	Not protected	Not protected	130	536
B	Fully protected	Fully protected	Fully protected	Not protected	150	692
C	Fully protected	Fully protected	Fully protected	Partially protected	158	824
D	Fully protected	Fully protected	Fully protected	Fully protected	202	909
E	Not protected	Not protected	Not protected	Not protected	138	579
F	Fully protected	Fully protected	Fully protected	Not protected	158	728
G	Fully protected	Fully protected	Fully protected	Partially protected	170	909
H	Fully protected	Fully protected	Fully protected	Fully protected	206	987

The efficiency of the SEC hamming code was tested by fault injection (Lima et al., 2001a). The results show that no errors were found in the application in presence of SEU. However this technique is not suitable for MBU. In (Lima et al., 2002a, 2002b), MBU were injected in the SEU tolerant 8051. The necessity of DEC hamming code and register refreshing in addition of the memory refreshing may be evident in next process technologies.

Maxwell (Maxwell, 2001) has a large range of SEU tolerant microprocessors protected by a patented radiation hardened RAD-PAK® technology that basically is a package shielding. The company offers microprocessors such as Intel 386, 486 and Pentium and SPARC from Sun. This same company also provides the microprocessor PowerPC from

Motorola with the CPU protected by TMR and the memory protected by EDAC. The TMR compares the output of each of 3 CPUs on a bit-by-bit basis. In the event of a single upset a simple voting scheme detects and selects the correct value. The advent of a second error would be uncorrectable, thus the processor is flushed and synchronized. In addition the components also have the package shielding.

Honeywell (Honeywell, 2003) offers fault-tolerant microprocessors based on device redundancy and EDAC techniques too. An example is the radiation hardened PowerPC 603, where the data and program memories are protected by SEC-DED hamming code and redundancy is applied in the internal registers. Aitech Defense Systems Inc. (Aitech, 2001) also provides a radiation tolerant PowerPC 750 protected by EDAC.

Lockheed Martin has developed a SEU tolerant PowerPC (G3) for JPL (JPL Laboratory, 2001). It provides a modular standard product that allows the spacecraft developer excellent flexibility in system configuration. There are over 800,000 storage elements in the PowerPC 750 (G3), all of which have been replaced with SEU hardened circuitry in the RAD750. The earlier RAD6000 employed resistor decoupling memory cells (figure 3-18) that require special polysilicon resistors in the manufacturing process. The RAM cells and latches in the RAD750 have been designed using hardening techniques for circuits that require no special process steps and optimize performance using the cells referred to in (Liu and Whitaker, 1992), figure 3-25. The memory and PROM located on the board have been protected by EDAC.

Atmel provides an 8-bit radiation tolerant micro-controller 80C32E, DSP microprocessor and a SPARC microprocessor for military and space applications (Atmel, 2001). The radiation tolerant DSP microprocessor Radiation from Atmel uses the Hit cell (Velazco et al., 1994), figure 3-20, in order to protect the memory cells against radiation. The Atmel SPARC microprocessor is protected by EDAC. The Atmel static RAM design separates the cells that represent the different data word bits. This feature virtually eliminates the risk of one impact provoking dual bit upsets (MBU) leaving only single bit upsets (SEU) that can be corrected by SEC hamming code. The additional processing associated with an EDAC protected solution is the initialization of the check bit RAM and a refresh procedure that performs read-write operations on the protected memory, also called scrubbing. The initialization of the check bit RAM does not introduce an overhead, since most space borne applications move their code from ROM to RAM at reset, and automatically initialize the check bit RAM at the same time. The scrubbing performed during processor idle time is necessary to eliminate the risk of two separate impacts generating a dual bit upset (MBU) in the same data word. However, if a dual bit upset in the same data word should occur it would still be detected and signaled by the EDAC, SEC-DED hamming code.

In (Gaisler, 2002), a fault-tolerant processor is proposed: the Spacelite, based on the SPARC V8 architecture. The techniques applied to this processor aim to detect and to tolerate one error in any on-chip register, and one error correction and double error detection in two adjacent bits in any on-chip memory structure (caches and tags). The approach to SEU fault-tolerance in the Spacelite processor is to divide all registers into two groups; primary and redundant. A primary register is defined as register carrying information, which is not present anywhere else in the system (processor or memory) and where an error in the register contents would cause a malfunction of the system. A redundant register is defined as a register that contains information that is replicated somewhere else in the system, and can be recreated by either reloading the register or performing other recovery actions. An error in a redundant register must also not alter the state or operation of the system in a way that will create a malfunction during the time it contains an erroneous value. To tolerate one random register error, all primary registers are designed fault-tolerant, either by replication or by use of error-correcting codes. The redundant registers need only be provided with error-detection functions, since they can be recovered from their redundant locations.

Individual fault-tolerant registers are implemented using TMR, three registers in parallel and a voter selecting the majority result. The benefit of such a scheme is that error masking and error-removal is implicit, and than no glitch is produced at the output when a SEU occurs. The register file is provided with a 32-bit single error correction (SEC) and double error detection (DED) EDAC instead of TMR cells to reduce the overhead. Errors in redundant registers are detected through parity generation and checking. Cache memories and tags are protected with two parity bits, one for odd and one for even data bits. This scheme makes it possible to detect a double-error in two adjacent bits. In case of an EDAC error, the corrected register value is written back to the register file when the instruction reaches the write stage, and the instruction is then restarted. An error in the cache memory (instruction or data) will automatically cause a cache miss, and the cache will be updated with the correct data from the main memory.

In (Rebaudengo et al., 2002), the software implemented fault tolerance (SIFT) is discussed to protect microprocessors against upsets in the sequential (SEU) and combinational logic (SET). Fault injection experiments have been performed to evaluate the capabilities of the SIFT technique of detecting transient faults in the internal memory elements of a processor and in its combinational logic. The originality of the strategy relies on the fact of being based on a set of simple transformation rules, which can be implemented on any high-level code. This reduces the costs for program hardening. The SIFT system implementations were tested under radiation in the 8051 micro-controller. Results show that SIFT tool was able to detect 88.2% of the upsets observed in the processor. However, this technique needs extra memory as the program code increases.

3.3 EXAMPLES OF SEU MITIGATION TECHNIQUES IN FPGAS

Field Programmable Gate Array (FPGA) devices are becoming increasingly popular with spacecraft electronic designers as they fill a critical niche between discrete logic devices and the mask programmed gate arrays. The devices are inherently flexible to meet multiple requirements and offer significant cost and schedule advantages. Since FPGAs are re-programmable, data can be sent after launch to correct errors or improve the performance of spacecraft.

The architecture of programmable logic components is based on an array of logic blocks that can be programmed by the interconnections to implement different designs. A FPGA logic block can be as simple as a small logic gate or as complex as clusters composed of many gates. The logic blocks of current commercial FPGAs are composed of one or more pairs of transistor, small gates, multiplexors, Lookup tables and and-or structures. The routing architecture incorporates wire segments of various lengths, which can be interconnected via electrically programmable switches. Several different programming technologies are used to implement the programmable switches. There are three types of such programmable switch technologies currently in use:

- *SRAM,* where the programmable switch is a pass transistor controlled by the state of a SRAM bit (SRAM based FPGAs)
- *Antifuse,* when an electrically programmable switch forms a low resistance path between two metal layers. (Antifuses based FPGAs)
- *EPROM, EEPROM or FLASH cell,* where the switch is a floating gate transistor that can be turned off by injecting charge onto the floating gate. These programmable logic circuits are called EPLDs or EEPLDs.

Customizations based on SRAM are volatile. This means that SRAM-based FPGAs can be reprogrammed as many times as necessary at the work site. The antifuse customizations are non-volatile and they can be pro-grammed just once. Each FPGA has a particular architecture. Programmable logic companies such as Xilinx and Actel offer radiation tolerant FPGA families. Each company uses different mitigation techniques to better take into account the architecture characteristics. Some companies from the space market are licensed to develop tolerant FPGAs, such as Aeroflex UTMC, which is licensed to QuickLogic, and Honeywell, which is licensed to Atmel. However, there is no current, finished space product based on the QuickLogic and Atmel FPGAs so far. Actel and Xilinx are the main commercial FPGA companies to share the market of space FPGAs nowadays as observed in the industry floor of the most important conferences of the area such as Military and Aerospace Applications of Programmable Devices and Technologies (MAPLD), Nuclear and Space Radiation Effect (NSREC), Radiation Effects on Components and Systems (RADECS) and Field Programmable Gate Array Symposium (FPGA).

The programmable logic devices are critically sensitive to SEU due to the large amount of memory elements located in these structures. Programmable logic devices must be strongly protected to avoid errors running in the space environment. There are two main ways to mitigate the radiation effects in Programmable Logic Devices: by high-level description or by architectural design.

Each method has a different implementation cost and it can be more suitable for some types of applications, FPGA topology and customization approach. For example, FPGAs programmed by antifuse topology are more like standard cell ASICs, as the customization cells (antifuse) are not susceptible to radiation effects. For this reason, techniques used in ASICs such as EDAC can be easily applied to the high-level description. At the architectural level, for instance, it is simple to replace all the flip-flops with hardened memory cells. As you will see later in this book, for FPGAs customizable by SRAM, applying high-level SEU mitigation techniques is not so simple because all the design blocks are sensitive to radiation. The same occurs when architecture design techniques are applied because of the FPGA matrix complexity.

3.3.1 Antifuse based FPGAs

The problem of SEU in antifuse FPGAs, more specifically based on the Actel architecture, has been addressed in (Katz et al., 1997, 1998, 1999; Wang et al., 2000). Actel started offering SEU tolerant FPGA families programmed by antifuse called SX some years ago (Actel, 2000). This family architecture is described as a "sea-of-modules" architecture, because the entire floor of the device is covered with a grid of logic modules with virtually no chip area lost to interconnect elements or routing. Actel's SX family has been improved in the past years. The first version provided two types of logic modules, identical to the standard Actel family, the register cell (R-cell) and the combinatorial cell (C-cell) exemplified in figure 3-31.

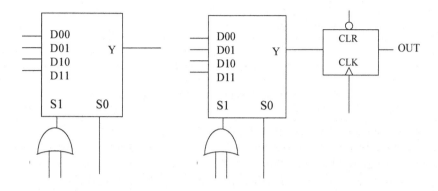

(a) Combinational ACT1 (C-cell) and sequential ACT1 (R-cell)

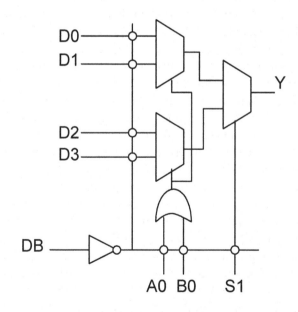

(b) C-cell represented in detail

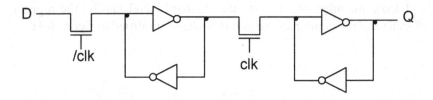

(c) R-cell: Latch representation

Figure 3-31. Architecture of Actel FPGAs

Interconnection between these logic modules is achieved using Actel's patented metal-to-metal programmable antifuse interconnect elements, which are embedded between the metal 2 and metal 3 layers. These antifuses are normally open circuit and, when programmed, form a permanent low-impedance connection. In this first SEU tolerant FPGA version (Actel, 2001), three proposed techniques for implementing the logic of the sequential elements in order to avoid upsets were presented: CC, TMR, or TMR_CC. The sequential elements are automatically implemented during the synthesis in the Symplify tool. The CC technique uses combinatorial cells with feedback instead of flip-flop or latch primitives to implement storage cells. For example, a DFP1, comprised of two combinational modules, would be used in place of a DF1. This technique can avoid SEU in CMOS technologies larger than 0.23 µm but it will not be able to avoid SEU in next-generation process technologies where the combinational logic can also be affected by charged particles. TMR is a register implementation technique where each register is implemented by three flip-flops or latches that "vote" to determine the state of the register. TMR_CC is also a triple-module-redundancy technique, where each voting register is composed of combinatorial cells with feedback (instead of flip-flop or latch primitives).

The CC flip-flops (CC-FFs) produce designs that are more resistant to SEU effects than designs that use the standard flip-flop (S-FF). CC-FFs typically use twice the area resources of S-FFs. Triple voting, or triple module redundancy (TMR), produces designs that are most resistant to SEU effects. Instead of a single flip-flop, triple voting uses three flip-flops leading to a majority gate voting circuit. This way, if one flip-flop is flipped to the wrong state, the other two would override it, and the correct value is propagated to the rest of the circuit. Because of the cost (three to four times the area and two times the delay required for S-FF implementations), triple voting is usually implemented using S-FFs. However, one can implement triple voting using only CC-FFs in the Synplify tool.

Currently, Actel offers the RTFXS and RTAXS FPGA families (Radiation Tolerant FX or AX architecture "Space" versions). These devices use metal-to-metal antifuse connections for configuration and include built-in TMR on all registers. These new SEU-hardened structures eliminate the need for TMR flip-flop designs implemented in HDL because the flip-flop is already protected by TMR at the architectural level (matrix). They use the D-type flip-flop proposed in (Katz et al., 2001; Wang et al., 2003), figure 3-32. Three D-type flip-flops are connected in parallel to the clock and data inputs. A voter (or majority circuit) is implemented by the top MUX to create a "hardened" output. The outputs of two flip-flops, A and B, go to the selects of the voter MUX. If both A and B read logic zero, MUX input D0 is selected. Since it is tied to GND, the output of the MUX will read logic zero. Similarly, if A and B read logic one, the output of the MUX will read logic one. If A and B disagree due to a SEU (or for other reasons), the MUX will select flip-flop C. We know C agrees with either A or B, and thus the MUX "voted" to produce data agreed on by two of the three flip-flops.

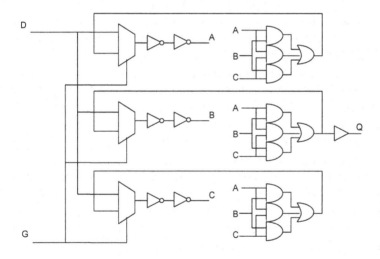

Figure 3-32. Built-in TMR on a memory cell in the Radiation Tolerant FX or AX architecture Actel's FPGAs

3.3.2 SRAM-based FPGAs

The SEU has a peculiar effect in SRAM-based FPGAs, as discussed in the previous chapter. As a consequence, it is not that simple to apply a high-level technique to this type of FPGAs, because all the implementation blocks (logic, customization and routing) are susceptible to upsets. Many solutions in the literature suggest new architecture topologies for SRAM-based FPGAs using hardened memory cells and innovative routing structures. Others solutions are high-level description techniques developed to be applied on the most popular family of SRAM-based FPGA, the Virtex® from Xilinx. The majority of the solutions for Virtex® are based on fault recovery, and they use partial reconfiguration and re-routing to correct upset and guarantee reliability. However, it is important to notice that many of the solutions that have been proposed for the Virtex® FPGA family in the high-level description are not very efficient because they do not take into account the peculiar effect of a SEU in the SRAM-based FPGA matrix, which is a permanent fault in the logic, customization and routing. Hardware redundancy is mandatory in this case to guarantee reliability.

3.3.2.1 SEU Mitigation Solution in high-level description

Xilinx has a military family for the Virtex® that is also used for space applications. Xilinx present many FPGA families for space applications. They are called QPro and it provides a commercial off-the-shelf system-level solution for aerospace and defense customers. It started with Virtex® QPRO family (Xilinx, 2000), which is fabricated on thin-epitaxial silicon wafers using the commercial mask set and the Xilinx 5-layer-metal 0.22 µm CMOS process. The use of epitaxial CMOS process technology has made Virtex® Single Event latchup immune (LET$_{th}$ >120 MeV*cm^2/mg, TID = 100 Krads(si)). Currently, Xilinx's Virtex®II QPro family, popular in space use, includes the same features as the regular Virtex®II Pro (integrated Power PC core, integrated SERDES, etc.) and is adapted for space use with extended temperature ranges, extended radiation tolerance, and ceramic packages.

In addition, Xilinx has proposed a high-level technique to mitigate SEU in the SRAM-based FPGA: the TMR approach in the high-level design description combined to reconfiguration (scrubbing) in order to avoid accumulation of upsets (Carmichael et al., 2000, 2001). This solution is complete to avoid single points of failure in the matrix as all blocks are triplicated. This solution has been investigated and experiment tests were performed. In the chapters 5, 6 and 7 the SEU mitigation technique based on TMR for the Virtex® FPGA from Xilinx is discussed in detail.

In (Alderighi et al., 2002), a design for a Xilinx FPGA-based multistage interconnection network (MIN) for a multi-sensor system that will be used in future scientific space missions is proposed. It is characterized by good concurrent fault diagnosis and fault detection capabilities. The fault tolerance strategy adopted is based on both network configuration and FPGA re-configuration. A slice control unit, one per each slice, allows changing the actual slice configuration, while the network control unit sets a new permutation. When a fault affecting a slice is detected, a finite-state machine fires and marks the actual configuration as faulty in the LUT with a fault. The state machine goes, in turn, to an active status and searches for an equivalent configuration available among those stored in the configuration LUT. When such a configuration is found, it is applied to fix the problem. To detect faults, a parity checker is used in each slice. Parity is actually the only invariant property that can be defined for the slice. The parity checker is endowed with self-checking ability, so that it can report faulty conditions relevant to the set of faults.

The limitation of this method is that only very few types of faults can be recovered by reconfiguring the network. All the faults in the customization routing in the FPGA that have permanent effect (as described in previous chapter) can only be corrected by FPGA reconfiguration (scrubbing). Results show that only 10% of the injected single upsets are recovered by the method, which are probably the faults in the LUTs. The majority are recovered by simple scrubbing.

3.3.2.2 SEU Mitigation Solutions at the Architectural level

In (Mavis et al., 1998), a FPGA has been developed for space and military applications based on a combination of four technologies: (1) radiation hardened nonvolatile SONOS (Silicon-Oxide Nitride-Oxide Semiconductor) EEPROM transistors, (2) unique SEU immune storage circuits, both for nonvolatile SONOS implementations and for volatile SRAM (static random access memory) implementations, (3) high-performance, radiation hardened, 0.8 μm, 3-level metal CMOS technology, and (4) new FPGA architectures developed specifically to accommodate good radiation-hardened circuit design practices. It is hardened for total ionization dose up to 200 krad(Si) and LET greater than 100 MeV-cm^2/mg. The NMOS SONOS transistors differ from conventional NMOS transistors in that the SONOS transistor has a variable threshold voltage while the NMOS transistor has a fixed threshold voltage. To erase a SONOS transistor (program it to a negative threshold voltage) a large (10 V) negative voltage is applied from the gate to the P-Well. This causes pair-hole tunneling into the nitride-oxide gate dielectric layer, and the resulting positive charge storage produces a depletion mode device. To store data in the transistor (program it to a positive threshold voltage) a large (10 V) positive voltage is applied from the gate to the P-Well. This causes electron tunneling into the gate dielectric and the resulting negative stored charge. In the SRAM version of the FPGA, volatile configuration storage is accomplished using a circuit derived from the DICE (dual interlocked storage cell) latch. The chip is programmed in much the same way as the SONOS version, using a shift register to serially load the row data and a column decode to select the column being written.

Actel has prototyped an SRAM-based FPGA (Wang et al., 1999). In this case, the standard SRAM memory cells were replaced by resistor-decoupling memory cells where the effectiveness depends on the resistor value; and DICE memory cells that are practically SEU immune at 0.25μm if only one node is hit. Figure 3-18 demonstrates the resistor decoupling memory cell and figure 3-23 the DICE cell, respectively. The resistor decoupling memory cell is able to avoid upsets because the resistors inserted in the feedback path work as filters to the transient pulse provoked by the charged particle. The DICE cell can avoid upsets because it stores data in two distinct parts, where if one part is corrupted the other one is isolated by the cell construction. However, conclusions presented in (Wang et al., 1999) show that multiple bit upsets (MBU) will limit both solutions in the future if the layout does not pay special attention to this issue. The redundancy hardening (DICE memory) is less effective than the resistor solution in two orders of magnitude regarding upset rates. The disadvantage of the resistor solution is temperature operation range sensitivity and the delay increase. The DICE also has a disadvantage in area overhead. It has 12 transistors compared to 6 transistors in the standard memory cell. For 0.18μm, the effectiveness of both solutions will be compromised and more ingenious designs will be needed in the circuit level.

Atmel (Atmel, 2001) also has published a version of an SRAM-based FPGA (AT6010) using the SOI process. The original logic block was not logically modified. The improvement achieved is limited to the SOI reliability in presence of SEU. Previous results have shown that the use of only SOI technology does not guarantee protection against SEU. Consequently, this solution from Atmel is not completely suitable for the space environment.

In (Kumar, 2003), a new SRAM-based FPGA is proposed based on the human immune system. This architecture adopts a distributed network without any centralized control. Error (antigen) detection is based on the principle of operation of the B-cell. Once an error is detected in a functional cell, a pre-determined spare cell replaces the functional cell by cloning its behavior. The proposed reconfiguration technique reduces the redundancy in the system.

Functional cells consist of a 10-bit control register, 1-bit error register and a logic block. The contents of the control register may be considered as the genetic code. The process of recognizing an error (antigen) by a B-cell is emulated in a functional cell by ensuring that the outputs generated are complementary. If the outputs are identical i.e. an error is present, the results are forced into high impedance. By forcing the outputs of a functional cell to 00 or 11 in the presence of an error, the role of a B cell is emulated. Once an error has been detected the 1-bit error register in the cell is set to 1, and all input information of the functional cell is loaded into the corresponding spare cell. The same occurs in the routing cell that also has a control register to detect the presence of faults. The authors did not go into much detail about faults in the control registers and how much time the system must be on hold until the logic is replaced by the spare logic.

3.3.2.3 Recovery technique

Many fault-tolerant approaches for SRAM-based FPGAs were presented in the past years related to re-routing and alternate configuration to avoid upsets in the used CLBs. The first problem of correcting faults by runtime reconfiguration without using any redundancy is the method to find the faults in the matrix. In (Mitra et al., 1998), a method that uses pseudo-exhaustive BIST is presented to detect upsets in the matrix. The technique has an extra advantage that it is not necessary to bring the whole system down while fault location is carried out. The problem is the time duration to detect faults. Some applications can not be on hold for a long time waiting for the system to be recovered.

An example of fault recovery based on reconfiguration and re-routing is shown in (Lach et al., 1998, 2000), where the physical design is partitioned into a set of tiles. The key element of this approach is partially reconfiguring the FPGA to an alternate configuration in response to a fault. If the new configuration implements the same function as the original, while avoiding the faulty hardware block, the system can be restarted. The challenging step is to identify an alternate configuration efficiently and to have fast runtime fault detection. In (Xu et al., 2000), another fault-tolerant approach for SRAM-Based FPGAs is presented related to the routing procedure. The problem is that both papers discuss radiation effects that are mainly upset (SEU). But in this case, the fault will be corrected in the next load of the bitstream (reconfiguration) and no work must be done in searching a new alternate configuration or routing. The methods are only justified if real permanent faults are present in the matrix due to total ionization dose, such as gate rupture, short or open metal wires.

In (Yu and Mccluskey, 2001), a solution to permanent fault repair in finer granularity of the FPGA is presented. A faulty module can be repaired by reconfiguring the chip so that a damaged configurable logic block (CLB) or routing resource is not used by the design. Many techniques have been presented to provide permanent fault removal for FPGAs through recon-figuration. One approach is to generate a new configuration after permanent faults are detected in computing systems. Another approach is to generate pre-compiled alternative FPGA configurations and store the configuration bit maps in non-volatile memory, so that when permanent faults are present, a new configuration can be chosen without the delay of re-routing and re-mapping. The authors propose some equivalent design candidates that can replace the original TMR design in case of a permanent fault, figure 3-33.

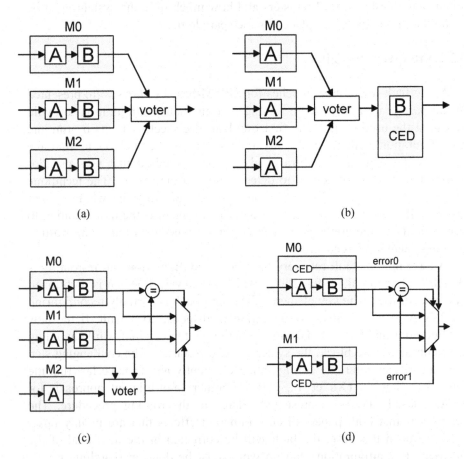

Figure 3-33. Design candidates modified from TMR. (a) The original TMR design. (b) A hybrid TMR-Simplex-CED design. (c) A duplex system with a checking block. (d) A duplex system with two CED blocks (YU and MCCLUSKEY, 2001)

For system transients, the authors suggest the use of traditional transient error recovery techniques. Typical examples include the roll-forward and rollback recovery techniques. Basically, these approaches are designed at the system level and thus are general to recover both Application Specific Integrated Circuit (ASIC) and FPGA systems. However, this assumption is not true, because in the SRAM-based FPGA not only can the logic be affected by upsets, but also the routing, which can invalidate the path to perform roll-forward and rollback techniques. The paper does not discuss the difference between real permanent faults (gate rupture, open or short metal wires) and the SEU that also has a permanent effect until the next reconfiguration.

In (Huang and Mccluskey, 2001), partial reconfiguration is also discussed to improve reliability by detecting and correcting errors in on-chip configuration data, but another problem is addressed in this paper: the memory coherence capability during partial reconfiguration. Because the LUTs can also implement memory modules for user applications, a memory coherence issue arises such that memory contents in user applications may be altered by the online configuration data recovery process. In this reference, the memory coherence problem is investigated, and it is proposed a memory coherence technique that does not impose extra constraints on the placement of memory-configured LUTs. Theoretical analyses and simulation results show that the proposed technique guarantees the memory coherence with a very small (on the order of 0.1%) execution time overhead in user applications. This technique is interesting and it can be further used with FPGA scrubbing in order to avoid SEU in the embedded memory too.

In summary, many fault-tolerant techniques have been proposed over the last years for SRAM-based FPGAs based on recovery, architectural design and high-level design. The majority of the techniques proposed in the past related to the high-level design method and the recovery procedure do not take into account all the details and effects of a SEU in the SRAM-based FPGA because this knowledge is very recent. No published paper before (Lima et al., 2003a) has established the difference between a real permanent fault and a SEU that has also a permanent effect in the LUTs, customization and routing cells in the FPGA. The next chapters show in detail the analysis of the effects of a SEU in the programmable matrix and the importance of using some kind of redundancy in order to ensure run time error recovery and scrubbing (continuous reconfiguration) to avoid accumulation of faults.

Chapter 4

ARCHITECTURAL SEU MITIGATION TECHNIQUES
For SRAM-based FPGAs

Programmable devices customizable by SRAM are composed of many components, such as complex logic blocks with lookup tables (LUTs), multiplexors and flip-flops, embedded memories, PLLs and dedicated routing segments, as explained in chapter 2. In addition, the present generation of FPGAs not only has the possibility of soft core insertion, but there are also hard microprocessor cores embedded in the chip to improve the data processing and performance, such as the family Virtex®II-Pro from Xilinx. Figure 4-1 diagrams a hypothetical topology.

The SEU mitigation problem in the next FPGA families with embedded hard processors can be analyzed in two parts: the microprocessor and the programmable logic. Consequently, each part can be sub divided into small logic blocks according to the functionality and some special features. Studies about the protection of a microprocessor have been done in the 8051-like micro-controller developed at UFRGS (CARRO; PEREIRA; SUZIM, 1996). All registers and the internal memory were protected by hamming code (LIMA et al., 2000a; LIMA et al., 2000b; COTA et al., 2000). The results have shown the reliability of this method and the necessity of refreshing in some parts of the circuit, mainly in the memory, in order to avoid accumulation of upsets. A fault injection system was developed (LIMA et al., 2001a) in order to test the standard and the full SEU tolerant 8051 in the presence of single and multiple upsets (LIMA et al., 2002a). Based on the references analysis in chapter 3 and the studies previously done, the problem of protecting microprocessors against SEU is relatively well understood, and the available techniques presented in the literature can be applied in order to achieve reliability.

73

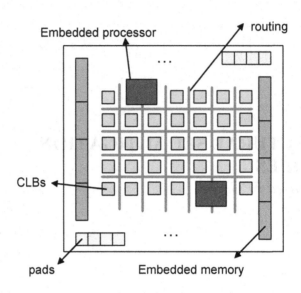

Figure 4-1. A case of Study: Hypothetical FPGA architecture

However, in the case of SRAM based FPGAs, the problem of finding an efficient technique in terms of area, performance and power is very challenging, because of the high complexity of the architecture. As previously mentioned, when an upset occurs in the user's combinational logic implemented in an FPGA, it provokes a very peculiar effect not commonly seen in ASICs. The SEU behavior is characterized as a transient effect, followed by a permanent effect. The upset can affect either the combinational logic or the routing. The consequences of this type of effect, a transient followed by a permanent fault, cannot be handled by the standard fault tolerant solutions used in ASICs, such as Error Detection and Correction Codes (EDAC), hamming code, or the standard TMR with a single voter, because a fault in the encoder or decoder logic or in the voter would invalidate the technique. Moreover, a fault in the routing of a SRAM-based FPGA can compromise the TMR protection as will be analyzed in the next chapters. The problem of protecting SRAM-based FPGAs against SEU is not yet solved and more studies are required to reduce the limitation of the methods currently used.

At previous chapter presented some architectural and high-level techniques for SRAM-based FPGAs. In this chapter, improvements to the architectural method will be addressed. The high-level method will be discussed in next chapters.

In the architectural level, the previous solutions leave open at least two problems to be solved:

- how to cope with SETs in the CLB logic to avoid upsets being stored in the flip-flop,

- how to cope with multiple bit upsets in the LUTs, routing and especially the embedded memory.

In this chapter, we propose the investigation and development of SEU mitigation techniques for SRAM-based FPGAs that can be applied to FPGAs with or without embedded processors, which can cope with the two problems still not solved. The SRAM based FPGAs were chosen because of their high applicability in space. Different than FPGAs programmed by antifuse that can be programmed just once, SRAM based FPGAs can be reprogrammed by the user as many times as necessary in a very short period. So, applications can be updated and corrected after launch. This feature is very valuable for space applications, because it can reduce the cost in update missions or even save missions that were launched with design problems.

First, it is necessary to analyze the amount of the sensitive area in the programmable matrix and their characteristics to propose improvements in the SEU mitigation techniques for SRAM-based FPGAs. Table 4-1 showed the set of configuration cells in a CLB tile of the Virtex® family. There are 864 memory bits responsible for the customization of the logic. Analyzing the percentage of each type of SRAM cell in the whole set of memory elements in the CLBs, the LUTs represent 7.4%, the flip-flops represent 0.46%, the customization bits in the CLB represent 6.36% and the general routing represents 82.9%.

Based on these results, the effect of an upset in the routing configuration (customization bits of the CLB and general routing) seems to be the major concern, totaling approximately 90% of the sensitive area in each CLB. This type of fault, as previously mentioned, has a permanent effect, which represents an open or short circuit in the final connections of the logic design. A solution that increases the area of this customization logic is not very attractive in final area and cost of this FPGA, since it means increasing a large portion of the FPGA, notice that routing corresponds roughly to 90% of the area of the FPGA.

In addition to these programmable cells presented in table 2-1, there are other memory elements in FPGA devices that can also be affected by SEU:

- SelectMAP (Selectable Microprocessor Access Port) latches.

- JTAG (Joint Test Action Group - IEEE Std. 1149.1x) TAP (Test Access Port) latches.

- Others latches of other built-in non-programmable feature, like bitstream uploading machine and power-on-reset (POR) control.

The main effects of a SEU in these latches are SEFI (Single Event Functional Interrupt) such as configuration circuit upsets and JTAG circuit upsets. There are few flip-flops or latches in the Power-on-Reset (POR) circuit, less than 40 latches or flip-flops, which leads to a very small cross-section. But they cannot be disregarded because an upset in one of these latches can force the chip to be reprogrammed. Figure 4-2 shows the location of these flip-flops in the FPGA matrix.

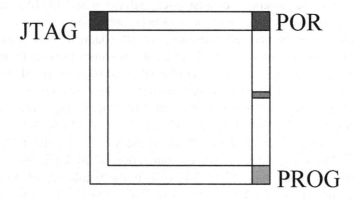

Figure 4-2. Special features elements in the SRAM-based FPGA matrix

Some solutions to protect the POR can be: TMR the whole block, replace the cells by SEU hardened memory cells or use extra logic to turn off the POR after the device is programmed by an external pin. In the next sections, some fault-tolerant techniques will be discussed to protect the SRAM cells of the LUT, flip-flops, routing and customization cells, and the embedded block RAM. The advantages and disadvantages of each technique were analyzed based on previous work results in ASICs implementations.

The first solution that can be studied is to replace some or all of the latches in the FPGA by SEU hardened flip-flops. Many hardened memory cells were designed during the last years. However, each one has different characteristics that can show more efficiency for some applications. Table 4-1 shows a summary of a comparison among them. The main characteristics used for the comparison are the number of transistor, the method, the SEU order effect, the ability to accumulate or not upsets and the SET immunity in combinational logic. For example, standard latches have a first order of susceptibility; in other words, they are upset by a single node strike. Some of them require multiple node strikes to upset cells such as TMR memory cells, DICE memory cell and simple temporal memory cells. Temporal latches built from DICE cells, for example, have a second and third order of susceptibility.

Table 4-1. Summary of Hardened Memory Cells: main Advantages and Drawbacks

Hardened memory cell	Method	I	II	III	IV	V	VI
Resistor memory cell	Decoupling resistor	8	Yes	Yes	No	No	No
IBM memory cell	Restore feedback	12	Yes	Yes	No	No	No
NASA I and II memory cell	Physical redundancy	12	Yes	Yes	No	No	No
DICE memory cell	Physical redundancy	12	Yes	Yes	No	No	No
HIT memory cell	Restore feedback	12	Yes	Yes	No	No	No
CANARIS memory cell	Restore feedback	32	Yes	Yes	No	No	No
TMR with one voter without refreshing	Physical redundancy	38	Yes	No	Yes	Yes	Yes *
TMR with three voters with refreshing	Physical redundancy	90	Yes	No	Yes	No	Yes **
Temporal memory cell	Temporal and physical redundancy	80	Yes	No	Yes	Yes	Yes
Temporal memory cell with DICE	Temporal and physical redundancy	134	Yes	Yes	Yes	No	Yes

I - Number of transistors
II - First order effect immunity to SEU and MBU
III - Second order effect immunity to SEU and MBU
IV - Third order effect immunity to SEU and MBU
V - Accumulation of upsets
VI - SET immunity

The hardened memory solution is suitable to replace the SRAM cells in the routing, general customization and lookup tables because they present a small overhead compared to logic redundancy technique and EDAC. Solutions such as IBM, NASA, DICE, HIT and resistor memory cells look interesting in the number of transistors and fault coverage. The final area will be around 2 times the original one, which is a very good result in terms of high-reliability.

For the LUT, for instance, if the cells are placed too close to each other, it is possible to use the solution of a TMR memory cell, where each cell is a DICE memory cell. In this case, this solution is robust to the 1st, 2nd and 3rd order of upsets. And because the LUT cells comprise only 7.4% of the cells, the impact in area will not be so intense. In (ROCKETT, 2001), a SEU immune memory cell based on decoupling resistors was developed for FPGAs. The design is asymmetric to provide that the data cell powers-up in a know state. In the paper, the size and the speed of the cell are discussed. The cells are not in the critical path, such as the cells that control de routing, for example, and hence do not need a high-speed operation and large transistors. In this case, the tolerance and the size are the main issue.

The embedded memory in the FPGA must be protected in order to avoid errors. EDAC is a suitable technique to correct upsets in memory structures, as previously discussed. An example is the hamming code that can be applied to embedded FPGA memory. However, as discussed in the previous chapter, hamming code is not able to cope with multiple upsets in the same coded word. And in the case of the embedded memory, it is very important to protect the cells against MBU for two main reasons:

-    new SRAM technologies (VDSM) are susceptible to MBU,

-    the scrubbing procedure does not reconfigure (update) the internal memory, consequently, upsets have a higher probability of accumulating in the memory.

So, a new code is needed to correct all possible double errors. The initial option would be using a Reed-Solomon code with capability to correct two different symbols. But this RS code has more than twice the area and delay overhead of the single symbol correction RS (HOUGHTON, 1997), which makes this solution inappropriate for hardware implementation in memory architectures. Previous work has been published on the use of RS code to protect memory (REDINBO; NAPOLITANO; ANDALEON, 1993), however it does not take into account double bit upsets in the same coded word, which is likely to occur in VDSM technologies.

An innovative solution has been developed able to correct all double bit upsets in VDSM memories. This solution combines hamming code and RS code with single symbol correction capability. This technique solves the problem of how to achieve 100% of double fault correction with a low-cost RS code. The hamming code protects the bits between the RS symbols. The number of bits protected by hamming will be the same as the number of symbols protected by Reed-Solomon, so this option does not significantly increases the area overhead. Figure 4-3 presents the insertion of hamming code in row already coded by RS code.

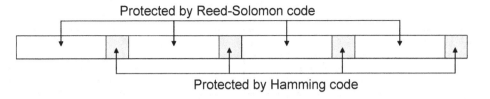

Figure 4-3. Schematic of a memory row protected by Reed-Solomon and hamming code

This solution is explained in detail in (NEUBERGER; LIMA; CARRO; REIS, 2003). Results show the efficiency of the proposed method in the presence of all single and double upsets and many types of multiple upsets. All double faults and a large combination of multiple faults were corrected by the method, faults type a, b, c, d, e, f, and g in figure 4-4. The only type of multiple faults that was detected but not corrected by the method is where multiple upsets (three or more) affect two different RS code symbols, fault type h in figure 4-4.

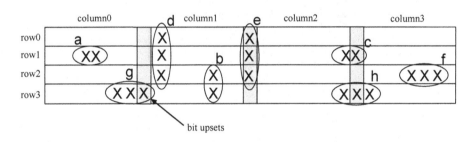

Figure 4-4. Schematic of a memory row protected by Reed-Solomon and hamming code

Figure 4-5 shows the final architecture of the double error tolerant memory. For instance, for 128-bit data protection, 14 extra bits are needed due to Reed-Solomon and 5 extra bits due to hamming code, totaling 19 parity bits for each data row. There are two encoder and decoder blocks, one for hamming code and another for RS code. The parity bits are also stored in the memory in a reserved area. The placement of all RS parity symbols and hamming parity bits must be also taken into account to avoid double upsets in the same hamming coded parity word or in two parity symbols of the same RS coded word. The RS encoder block can be adapted to any size of data memory as presented in (Neuberger; Lima; Carro; Reis, 2003).

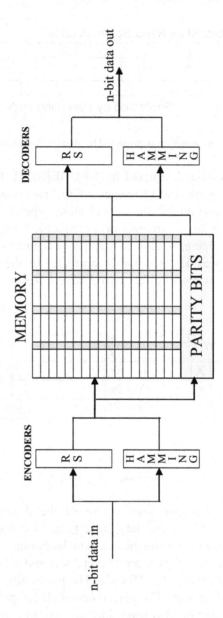

Figure 4-5. Hamming and RS code in memory architecture

The case study memory was described in VHDL and prototyped in a Virtex®E FPGA using BlockRAMs and CLBs in order to be evaluated in terms of area, performance and fault coverage. Results are presented in table 4-2.

Table 4-2. Area and Delay of Reed-Solomon and hamming codes used to protect a memory

	16-bit Hamming		112-bit RS	
	Encoder	Decoder	Encoder	Decoder
# 4-LUTs	22	99	215	538
# Extra ffs	5 x # of rows		14 x # of rows	
Delay (ns)	9.3	21.7	14.5	47.6

Results show that the fault tolerant memory has an area overhead that is basically the area used by the encoder and decoder blocks. Only two more BlockRAMs are needed, one to store the RS redundancy symbols and other to store the hamming extra bits. The performance penalty in the fault tolerant memory synthesized in the FPGA is around 50%. This penalty can be reduced when the encoder and decoder blocks are implemented using random logic instead of the CLBs (prototype version).

In summary, the proposed method that combines RS code and hamming code to protect memory against SEU is an attractive fault-tolerant technique to be applied in the new hardened SRAM-based FPGA. It is able to protect the memory against all double faults and a large set of multiple faults. It does not present a large impact in area and it does not interfere with the normal operation and customization of the current embedded memory cell. The presented method is innovative. In the literature, only one approach has been found similar to this one (Redinbo; Napolitano; Andaleon, 1993), but it uses only Reed-Solomon code and does not correct double faults at the interface of two different symbols. Our approach corrects this type of fault, avoiding the choice between correcting one symbol error and detecting double symbol errors.

The triple modular redundancy (TMR) is another SEU mitigation technique. There are many TMR topologies. Each implementation is associated with different area penalties and fault coverage. The system requirements and the architecture must be analyzed in order to correctly choose the most convenient approach. Table 4-3 shows a summary of the main approaches of TMR.

The CLB flip-flops receive the output of the multiplexors that set up the signal path from the LUT in the CLB slice. If a transient fault (SET) occurs in one of the multiplexors, this upset must not be stored in the flip-flops. Consequently, it is not sufficiently reliable to replace the CLB flip-flop by a hardened flip-flop. It is also necessary to insert some kind of fault detection and correction in the input of this flip-flop to filter SETs. The combinational logic does not need to be changed. A possible solution is to combine the temporal latch composed of DICE memory cells, presented in (Mavis; Eaton, 2000; Mavis; Eaton, 2002) with the TMR approach with refreshing. The final flip-flop shows a high reliability to 1st, 2nd and 3rd order of SEU and SETs, refreshing of SEU and additionally a small impact in the final area because the flip-flops correspond to less than 1% of the total sensitive area. Figure 4-6 shows this hardened flip-flop topology.

Table 4-3. Summary of TMR approaches: main Advantages and Drawbacks

TMR Versions	I	II	II	IV	V
TMR Device without refreshing	Yes	Yes	Yes	Yes	Yes*
TMR in sequential logic without refreshing	See table 4-1,			Yes	No
TMR in sequential with refreshing	according to the			No	No
	TMR latch				
TMR combinational and sequential logic without refreshing	Yes	No	Yes	Yes	Yes
TMR combinational and sequential logic with refreshing	Yes	No	Yes	No	Yes

I - First order effect immunity to SEU and MBU
II - Second order effect immunity to SEU and MBU
III - Third order effect immunity to SEU and MBU
IV - Accumulation of upsets
V - SET immunity

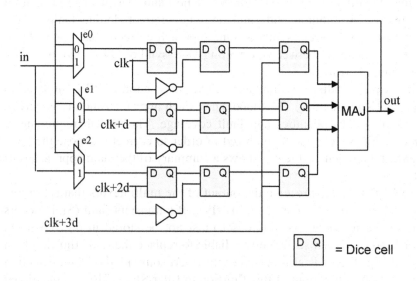

Figure 4-6. Proposed SEU and SET Hardened flip-flop with refreshing

Chapter 5

HIGH-LEVEL SEU MITIGATION TECHNIQUES
For SRAM-based FPGAs

The previous chapter discussed fault-tolerant techniques in the architectural level for SRAM-based FPGAs. Although these solutions can achieve a high reliability, they also present a high cost, because since they change the matrix, they need investment in development, test and fabrication. So far, there are very few FPGA companies that are investing in designing fault-tolerant FPGAs as this market is still focused in only military and space application, which is a very small market compared to the commercial one. However, because of the technology evolution, applications at the atmosphere and at ground level have been starting to face the effect of neutrons, as mentioned in chapter 2. As a result, fault-tolerant techniques begin to be necessary in many commercial applications that need some level of reliability.

A less expensive solution is a high-level SEU tolerant technique that can be easily implemented by the user or by the company designers in commercial FPGAs or in parts manufactured by a technology that can avoid latch up and reduce the total ionization dose, as the Virtex® QPRO family (Xilinx, 2000). The high-level SEU mitigation technique used nowadays to protect designs synthesized in the Virtex® architecture is mostly based on TMR combined with scrubbing (Carmichael; Caffrey; Salazar, 2000; Carmichael, 2001). The TMR mitigation scheme uses three identical logic circuits (redundant block 0, redundant block 1 and redundant block 2), synthesized in the FPGA, performing the same task in tandem, with corresponding outputs being compared through a majority vote circuit. The TMR technique for Virtex® is presented in details in (Carmichael, 2001), and more examples are also presented in (Lima et al., 2001b).

5.1 TRIPLE MODULAR REDUNDANCY TECHNIQUE FOR FPGAS

The correct implementation of TMR circuitry within the Virtex® architecture depends on the type of data structure to be mitigated. The logic may be grouped into four different structure types: Throughput Logic, State-machine Logic, I/O Logic, and Special Features (Select block RAM, DLLs, etc.). The throughput logic is a logic module of any size or functionality, synchronous or asynchronous, where all of the logic paths flow from the inputs to the outputs of the module without ever forming a logic loop. In this case, it is necessary to just triplicate the logic, creating three redundant logic parts (0, 1 and 2). No voters are required, as the FPGA output will be by default voted later.

The state-machine logic is any structure where a registered output, at any register stage within the module, is fed back into any prior stage within the module, forming a registered logic loop. This structure is used in accumulators, counters, or any custom state-machine or state-sequencer where the given state of the internal registers is dependent on its own previous state. In this case, it is necessary to triplicate the logic and to have majority voters at the outputs. The register cannot be locked in a wrong value, for this reason there is a voter for each redundant logic part in the feedback path making the system able to recover by itself. Figure 5-1 shows a general example of this structure.

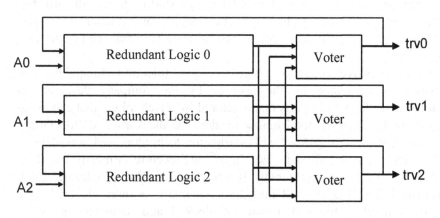

Figure 5-1. TMR Logic with Voter

The majority voter, figure 5-2 (a), can be easily implemented by one LUT. Because the LUT can be upset, the voters are also triplicated. In this way, if one voter is upset, there are still two voters working properly. For designs constrained by available logic resources, the majority voter can be implemented using the Virtex® 3-state buffers instead of LUTs, Figure 5-2 (b). There are two 3-state buffers per CLB. Figure 5-2 (c) shows the 3-state buffer schematic in the Virtex® matrix.

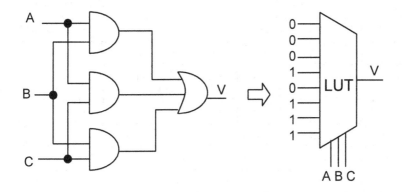

(a) 3-Input Majority Voter Schematic

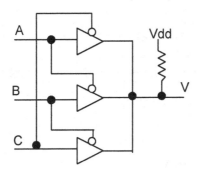

(b) 3-Input Majority Voter Implemented by 3-State Buffers

Figure 5-2. Majority Voters (CARMICHAEL, 2001)

The primary purpose of using a TMR design methodology is to remove all single points of failure from the design. This begins with the FPGA inputs. If a single input was connected to all three redundant logic legs within the FPGA, then a failure at that input would cause these errors to propagate through all the redundancies, and thus the error would not be mitigated. Therefore, each redundant leg of the design that uses FPGA inputs should have its own set of inputs. Thus, if one of the inputs suffers a failure, it will only affect one of the redundant logic parts. The outputs are the key to the overall TMR strategy. Since the full triple module redundancy generates every logic path in triplicate, there must ultimately be a method for bringing these triple logic paths back to a single path that does not create a single point of failure. This can be accomplished with TMR outputs majority voters inside the output logic block, as presented in figure 5-3.

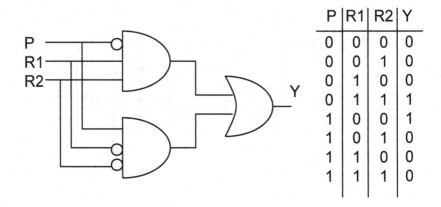

P	R1	R2	Y
0	0	0	0
0	0	1	0
0	1	0	0
0	1	1	1
1	0	0	1
1	0	1	0
1	1	0	0
1	1	1	0

(a) Output Logic Voter and the Truth Table

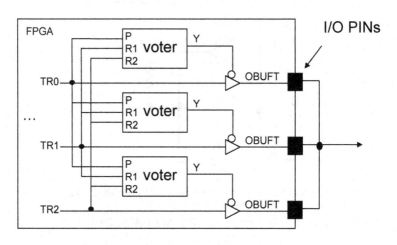

(b) The TMR Output Voter Scheme
Figure 5-3. Voter Structure in the Virtex® Output Logic (CARMICHAEL, 2001)

The Virtex® architecture provides a number of special features, such as block RAM (BRAM), DLLs, etc, which require specialized methods for implementing effective redundancy. A reliable method to TMR the BRAM is to constantly refresh the BRAM contents, figure 5-4. Since these are dual port memories, one of the ports could be dedicated to error detection and correction. But this also means that the BRAM could only be used as single port memories by the rest of the user logic. To refresh the memory contents, a counter may be used to cycle through the memory addresses incrementing the address once every n clock cycles. The data content of each address is voted at a determined frequency and the majority voter value written back into the cells.

A typical FPGA design will be implemented with signals that were resolved to a logic constant (VCC or GND), but could not be entirely optimized out of the design. When the Place and Route (PAR) tools deal with the VCC and GND signals, they are implemented in a way that maximizes device resource utilization. This is accomplished by utilizing "Keeper" circuits that exist at the input pins of all CLBs and I/O blocks (IOBs). Keepers lie in series with routing channels and logic block input pins. When the routing channel carries an active signal, the keeper is transparent. But when the channel is unused, the keeper will keep its last known value - which was determined when the device was initially powered-up or re-initialized by activating the FPGA input PROG. When a logic element (i.e. flip-flop) inside a logic block (i.e. CLB or IOB) requires a logical constant, such as a VCC or GND, this logical constant may be obtained from the keeper circuit of an unused pin of the logic block. Its polarity may be selected by programmable inversion within the logic block.

An SEU may upset, or alter, the state of a keeper circuit either by direct ionization, or indirectly by momentarily connecting an active routing channel to the input of the keeper. In either case, the result is a functional disturbance that can neither be detected by readback nor corrected by partial reconfiguration. Therefore, this type of error is known as a "persistent error", and it can only be corrected by completely re-initializing the FPGA. Schematic designers should be careful to examine the primitive implementation of all library macros that are likely to contain registers, before using them in their design. Even if the macro provides clock enable and reset pins at the top level, the primitive implementation may be different than expected. Similarly, if a VHDL user describes a synchronous process without specifying a clock-enable or initialization function, the synthesis tool will implement this function by using primitives and connecting all unused pins to the correct logical constant, thus creating VCC and GND. In order to avoid persistent errors, user VCCs, user GNDs and user clock enables for each redundant logic part must be created in the design as inputs.

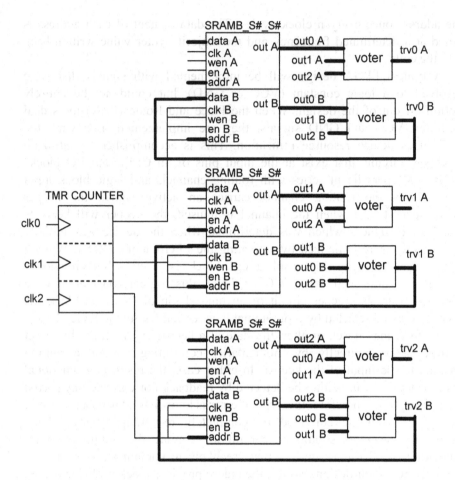

Figure 5-4. BRAM TMR with Refreshing (CARMICHAEL, 2001)

5.2 SCRUBBING

The use of TMR in the design is not sufficient to ensure reliability for a long period of time, as upsets can accumulate in the matrix, provoking an error in the TMR. Note, as explained in chapter 2, that the upsets located at LUTs and in the routing configuration cells will not be removed until the next configuration of the device. Consequently, it is necessary to clean up all the upsets in such a frequency as to guarantee the correct functionality of the TMR methodology. The first technique proposed to clean the upsets inside the matrix was based on readback of the bitstream, detecting an upset and reloading the original one (Carmichael; Caffrey; Salazar, 2000). The problem of this technique is that it is too time consuming.

A simpler method of SEU correction is to omit readback and detection of SEUs and simply reload the entire CLB Frame segment at a chosen interval (Xilinx, 2000c). This is called "scrubbing". Scrubbing requires substantially fewer overheads in the system, but does mean that the configuration logic is likely to be in "write mode" for a greater percentage of time. However, the cycle time for a complete scrub can be made relatively short. The scrubbing allows a system to repair SEUs in the configuration memory without disrupting its operations. It is performed through the Virtex® SelectMAP interface. When the FPGA is in this mode, an external oscillator generates the configuration clock that drives the PROM and the FPGA. At each clock cycle new data is available on the PROM data pins. One example is the Flash-PROM XQR18V04, which provides a parallel frequency up to 264 Mbps at 33 MHz. Figure 5-5 shows the scrubbing scheme.

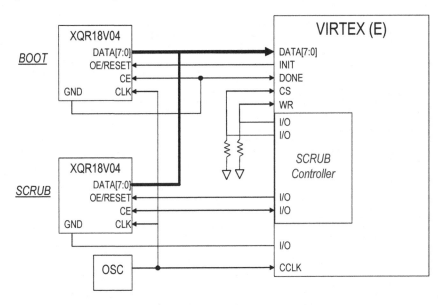

Figure 5-5. Scrubbing configuration scheme

The scrubbing cycle time depends on the configuration clock frequency and on the readback bitstream size. For the XQVR300, it is necessary to utilize 207,972 clock cycles in order to perform the full scrubbing load (Scrub cycle = # clock cycles x clock period). The scrubbing rate describes how often a scrubbing cycle must occur. It is determined by the expected upset rate of the device for the given application. Upset rates are calculated from the static bit cross-section of the device and the charged particle flux the application or mission is expected to endure. The scrubbing rate should be set such that any SEU on the configuration memory will be fixed before the next upset will occur. In reality the scrubbing rate is minimized to be equal to the scrubbing cycle. In this way, configuration logic is always being refreshed. The implemented design can also have influence in the selection of the scrubbing rate. A good "rule of thumb" is to place the scrubbing rate one order of magnitude or more above the expected upset rate. In other words, the system should scrub, on the average, at least ten times between upsets.

Chapter 6

TRIPLE MODULAR REDUNDANCY (TMR) ROBUSTNESS
For SRAM-based FPGAs

The robustness of the TMR technique implemented in a high-level description language synthesized in the Virtex® FPGA was evaluated by injecting faults in the configuration bits of the matrix (LUTs and configuration routing cells) and in the presence of protons generated by an electronic beam in a radiation ground test facility. In order to better understand how the faults were injected in the configuration cells and how they will effect the design operation, it is first necessary to study the configuration memory of the FPGA.

The Virtex® configuration memory can be visualized as a rectangular array of bits that is called the bitstream, figure 6-1. The configuration memory array is divided into three separate segments: The "CLB Frames", "BRAM0 Frames" and "BRAM1 Frames". The two BRAM segments contain only the RAM content cells for the Block RAM elements. The BRAM segments are addressed separately from the CLB Array. Therefore, accessing the Block RAM content data requires a separate read/write operation. Read/Write operations to the BRAM segments should be avoided during post-configuration operations, as this may disrupt user operation.

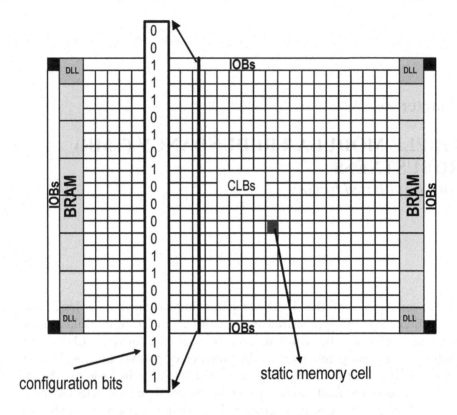

Figure 6-1. Virtex® Architecture Overview

The CLB Frames contain all configuration data for all programmable elements within the FPGA. These include all Lookup Table (LUT) values, CLB, IOB, and BRAM control elements, and all interconnect control. Therefore, every programmable element within the FPGA can be addressed with a single read or write operation. These entire configuration latches can be accessed without any disruption to the functioning user design, as long as LUTs are not used as distributed RAM components.

While CLB flip-flops do have programmable features that are selected by configuration latches, the flip-flop registers themselves are separate from configuration latches, and cannot be accessed through configuration. Therefore, readback and partial configuration will not affect the data stored in these registers. However, when a LUT is used as either a distributed RAM element or as a shift register function, the 16 configuration latches that normally only contain the static LUT values are now dynamic design elements in the user design. Therefore, the use of partial reconfiguration in a design that contains either LUT-RAM (i.e., RAM16X1S) or LUT-Shift-register (SRL16) components may have a disruptive effect on the user operation. For this reason the use of these components cannot be supported for this type of operation.

However, Select block RAMs (BRAM) may be used in such an application. Since all of the programmable control elements for the BRAM are contained within the CLB Frames and the BRAM content is in separate frame segments, partial reconfiguration may be used without disrupting user operation of the BRAM as design elements.

The configuration memory segments are further divided into columns of data frames. A data frame is the smallest portion of configuration data, which may be read from, or written to, the configuration memory. The bits are grouped into vertical frames that are one-bit wide and extend from the top to the bottom of the array composing a column defined by a major address (Xilinx, 2000c). Each matrix column is associated with a major address and to a different number of frames according to the nature of the column, shown in table 6-1.

The frames are read and written sequentially with ascending addresses for each operation. The frame size depends on the number of rows in the device. The number of configuration bits in a frame is 18 x (# of CLB rows +2), and is padded with zeros on the right bottom (LSB) to fit a 32-bit word.

The frame organization differs for each type of column. Each frame is located vertically in the device with the front of the frame at the top. Table 6-2 shows the CLB column frame, IOB column frame and BRAM content organization. The frame top is showed on the left.

The CLB tile is composed of the CLB logic and the surrounding interconnection that is placed in a determined row and column in the matrix. There are 864 customization bits per CLB tile distributed in 48 frames with 18 bits each, figure 6-2. The bits can be divided in Look-up table bits (7.4%), CLB configuration bits (6.8%), interconnection bits (84.2%) and 3-state buffer configuration bits (1.6%).

Table 6-1. Virtex® Configuration Column Type

Column Type	# of frames	# per device
center	8	1
CLB	48	# CLB columns
IOB	54	2
BRAM interconnect	27	# of blocks SelectRAM columns
BRAM content	64	# of blocks SelectRAM columns

Table 6-2. Frame Organization

CLB column frame				
Top 2 IOB	CLB R1	...	CLB Rn	Bottom 2 IOB
18 bits	18 bits	...	18 bits	18 bits
IOB column frame				
Top 3 IOB	3 IOBs	...	3 IOBs	Bottom 3 IOB
18 bits	18 bits	...	18 bits	18 bits
Block SelectRAM content				
PAD	RAM R0	...	RAM RN	PAD
18 bits	72 bits	...	72 bits	18 bits

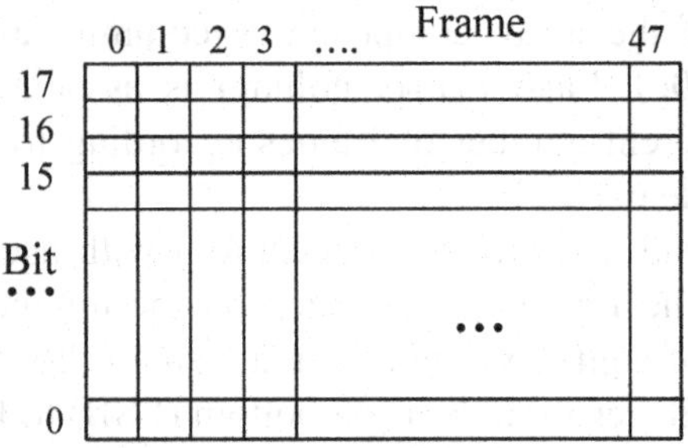

Figure 6-2. CLB Tile Map

The fault injection investigation is an efficient method to qualify the design before submit the part under radiation. The fault injection in SRAM-based FPGAs is defined as a bit flip in all bits of the configuration bitstream. In this way, it is possible to evaluate the effects of an upset in all sensitive areas of the programmable matrix. Some of these bits are directly related to the user's design combinational and sequential logic, and some of them are related to the FPGA architecture and design implementation.

The fault injection analysis was executed in four main steps to evaluate a first TMR design experiment. First, a fault injection tool was used to catalogue all the configuration bit locations that caused a dynamic error in the TMR design. Then all the reported bits were identified in the general FPGA matrix in terms of row, column and functionality. Based on this information, it was possible to identify those bits in the FPGA IC schematics by the use of JBits classes (Guccione et al., 1999). The third step identified the correlation between the bit location in the FPGA IC schematic and its location in the design under test in the FPGA editor tool. The last step was the characterization of the error.

6.1 TEST DESIGN METHODOLOGY

The TMR test design methodology used to analyze the SEU in the Virtex® FPGA consists of a TMR counter design replicated in the circuit in order to fill the resources of the part (XQVR300). All CLBs were used to implement eight TMR 32-bit counters with pipeline design. The design can be divided in three groups: the redundant logic part 0, redundant logic part 1 and redundant logic part 2. Each redundant group is composed of eight 32-bit counters. Figure 6-3 illustrates the design scheme.

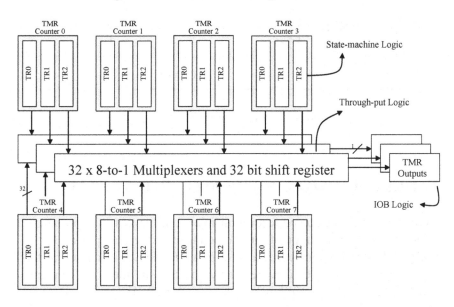

Figure 6-3. TMR design of a 32-bit pipelined counter

In order to detect an error in one of the 32-bit counters, the eight 32-bit counters located in the same redundant logic group are compared against each other. There is one comparator for each group. Comparators 0, 1 and 2 report an error in the redundant part 0, 1 and 2 respectively.

The three redundant logic groups are finally compared in the majority voter located in the output logic block. The error flag, a result of the majority voter, reports if there is an error in two or more redundant parts. A schematic of this approach is illustrated in figure 6-4.

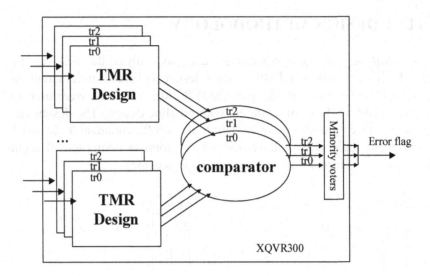

Figure 6-4. TMR Design Methodology

6.2 FAULT INJECTION IN THE FPGA BITSTREAM

There are many ways to perform fault injection. First, it is important to have a good model for the type of fault to be injected, for example, a SEU in a memory cell can be modeled as a bit-flip, as previously discussed in the early chapters. So, a SEU can be easily injected in memory elements described in hardware description language, such as VHDL or Verilog, by performing a XOR operation between the original stored value and a mask, which defines the bit to be flipped. In the case of SRAM-based FPGAs, the bit-flip can occur in any bit of the bitstream, as a result, it is natural to perform the fault injection directly in the FPGA bitstream without deal with the signals design and topology in the hardware description.

The JBits SDK (Software Development Kit) (Guccione et al., 1999) is a set of Java classes which provide an Application Program Interface (API) into the Xilinx Virtex FPGA family bitstream. This interface operates on either bitstreams generated by Xilinx design tools, or on bitstreams read back from actual hardware. This provides the capability of designing, modifying and dynamically modifying circuits in Virtex devices. The programming model used by JBits is a two dimensional array of CLBs.

The JBits API gives the user the ability to configure CLBs directly. That can be done by exploring the com.xilinx.JBits.Virtex.Bits classes, where all classes to access the resources of the CLBs are included. By using JBits classes, it is possible to perform a selective fault injection in the bitstream, which can also reduce the time spent in fault injection (Kinzel, et al., 2005). Figure 6-5 presents the fault injection and analysis flow by using JBits.

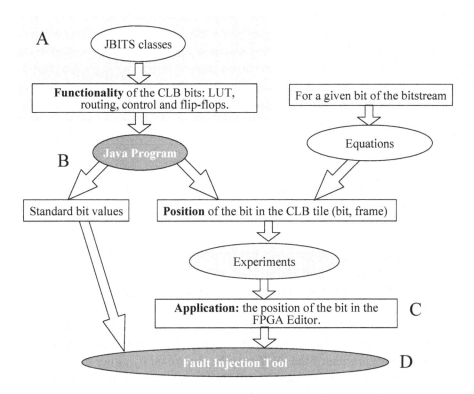

Figure 6-5. Bitstream Fault Injection and Analysis Flow

In the next paragraphs, a fault injection experiment using a fault injection tool is presented. This fault injection tool is able to corrupt all the bits of Virtex® bitstream in a sequential way, or individually by choosing a specific bit location. The objective of this tool is to analyze the effect of a single bit upset in a TMR design implemented in the Virtex® architecture. All single bit upsets able to cause an error in the TMR design were cataloged for investigation. In this text, this tool will be called Virtex® fault injection tool.

In principle, no single bit upset in the bitstream should cause an error in the TMR design, if a single upset error affects only one redundant part of the design. By TMR definition, if one redundant part is corrupted by an upset, the majority voters continue voting the correct value from the two other redundant uncorrupted parts.

The Virtex® fault injection tool can upset a single bit in the bitstream, starting from a user defined sequentially major address and frame, or it can upset one specific bit when the user defines the major address, frame, frame byte and bit. Fault injection is performed in three steps, presented in table 6-3. The three-step method guarantees no double upsets for any short period of time.

Table 6-3. VIRTEX® Configuration Column Type

Fault injection steps	Bitstream example
Read the bitstream:	... 0110010101010...
Corrupt one bit and load the bitstream:	... 1110010101010...
Correct the previous bit and load the bitstream:	... 0110010101010...
Reset the flip-flops	... 0110010101010...

Each time an error is reported by the test design comparator, the fault injection tool shows the location of the upset bit that caused the error. The tool reports the major address, frame, frame byte and bit location. Using this information it is possible to know the exact location of the bit in the bitstream, and as consequence in the FPGA matrix.

The fault injection test platform, shown in figure 6-6, is made from two AFX V300PQ240-100 daughter cards; a MultiLINX cable used as an interface to a host PC, and a control panel. The system can operate stand-alone or in conjunction with a host PC and test software. The control panel communicates directly with the control chip to specify the mode of operation. Configuration of the DUT may be controlled by either the control chip or the test software via the MultiLINX Cable. The control chip also controls the dynamic operation of the DUT and dynamic error detection.

6.3 LOCATING THE UPSET IN THE DESIGN FLOORPLANNING

For a given bit located in the bitstream in the CLB frames, there is a unique address location that is defined by the Major Address, Frame, Frame Byte and Bit position. In order to know the location of this bit in the FPGA matrix and consequently its purpose in the user design, it is necessary to follow the next steps.

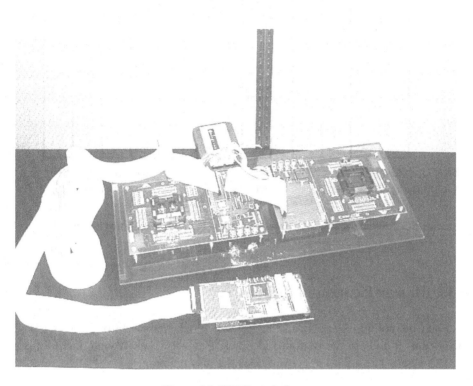

Figure 6-6. SEU Test platform

6.3.1 Bit column location in the matrix

The CLB address space begins with '0' for the center column and alternates between the right and left halves of the device for all the CLB columns, then IOB columns, and BRAM interconnect columns, as illustrated in figure 6-7. Analyzing the figure, if the major address is 23, for example, it means that the CLB column is 24. More information can be obtained in the application note (Xilinx, 2000c).

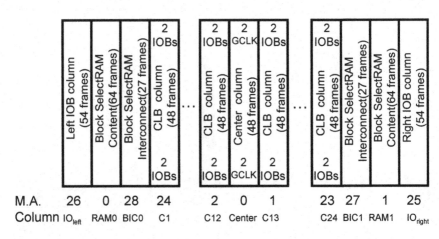

Figure 6-7. Example of Major Address organization in Virtex® family and it respective number of frames

6.3.2 Bit row location in the matrix

Each bit column starts from the top I/O block, passes through all CLB blocks and it ends in the bottom I/O block. Each row has 18 bits, including the I/O block. The equation 6.1 provides the row position.

Row = (Byte Frame x 8 + bit) / 18

The frame, frame byte and bit data are used to obtain the exact bit location in the CLB tile by using the equation 6.2.

CLB tile bit = 17 - [Byte Frame x 8 - Floor((Byte Frame x 8 + Bit) / 18), 18) + Bit]

6.3.3 Bit location in the CLB

Each bit of the CLB tile has been identified in the FPGA IC schematic by using JBits classes and therefore in the design floorplanning by using ISE Xilinx tool. In this way it was possible to build a design flow from the upset bit information coming from the fault injection tool (major address, frame, frame byte and bit) and the final design floorplanning bit location.

6.3.4 Bit Classification

The CLB map has the general bit classification (LUT, flip-flop, customization or routing) in terms of frame number and the bit location in the CLB. This map gives just a general view. In order to be able to find the specific location of the bit in the CLB architecture a table containing all the bit names is used. Table 6-4 shows a portion of the used CLB map table containing just the frame address 0.

Table 6-4. Bit Classification in the CLB

Frame	Bit	Function
0	0	Hex mux
0	1	Hex mux
0	2	Single
0	3	Single
0	4	Single
0	5	Single
0	6	Single
0	7	Single
0	8	Single
0	9	Imux
0	10	Imux
0	11	Imux
0	12	Imux
0	13	Slice
0	14	Slice
0	15	Slice
0	16	Slice
0	17	Omux

Each frame and bit location can be associated to a CLB functionality position, which the correlation is informed by JBits. For example, a routing position such as singles. Iw22o7 shows the connection between the single wire West 22 and single wire out 7.

An intense search was done in the Virtex® JBits in order to associate each name to each structure. But still this was not enough to enable one to associate the bits in the FPGA architecture to the signals in the user design. The software package called XDL (Xilinx, 2001b) must be used to see all the connections and instantiations in the user design. XDL is a full featured design language that provides direct read and write access to Xilinx proprietary Native Circuit Description (ncd), figure 6-8. It is a single tool with 3 fundamental modes: report device resource information, convert NCD to XDL (ncd2xdl) and convert XDL to NCD (xdl2ncd).

```
   Command example: xdl -ncd2xdl design.ncd design.xdl


inst "counter6/I$2/pipe0/N$3(3)"  "SLICE",    placed R31C7
CLB_R31C7.S1,
 cfg "CKINV::1 DYMUX::1 DXMUX::1
   F:counter6/I$2/pipe0/stage2/pipe3/I$8:#LUT:D=(~A2*A3)
   G:counter6/I$2/pipe0/stage2/pipe2/I$8:#LUT:D=(~A2*A3)
   CEMUX::CE_B SRMUX::SR_B GYMUX::G FXMUX::F
SYNC_ATTR::SYNC
   SRFFMUX::0 INITY::LOW
   FFX:counter6/I$2/pipe0/stage2/pipe3/I$3:#FF
   FFY:counter6/I$2/pipe0/stage2/pipe2/I$3:#FF
INITX::LOW

_PINMAP:24:0,1,2,3,4,5,8,6,7,9,10,11,14,12,13,15,16,17,
18,19,20,21,22,23";


net "count_data_tr2_0(1)",
 outpin "count_data_tr2_0(0)" YQ,
 inpin "mux2/L1.L1.1_mux/G_22_73" G4,
 pip R32C46 S0_YQ -> OUT0,
 pip R32C46 OUT0 -> N0,
 pip R31C46 S0 == E22,
 pip R31C47 W22 -> W_P22,
 pip R31C47 W_P22 -> S1_G_B4,
 # net "count_data_tr2_0(1)" loads=1 drivers=1 pips=5
rtpips=0;

net "L0.L0.3_C/Pipe2/pipeline_5(0)",
 outpin "L0.L0.3_C/Pipe2/pipeline_5(1)" YQ,
 inpin "count_data_tr2_3(0)" F2,
 pip R16C46 S1_YQ -> OUT5,
 pip R16C46 OUT5 -> E15,
 pip R16C46 E15 -> E_P15,
 pip R16C46 E_P15 -> S0_F_B2,
 # net "L0.L0.3_C/Pipe2/pipeline_5(0)" loads=1 drivers=1
pips=4 rtpips=0;
```

Figure 6-8. Example of design connection file (.ncd)

Based on the information of the Virtex® architecture and the equations presented in this section, i is possible to make a correlation between the programmable bit in the bitstream and its location in the design floorplanning that can be observed in the ISE software. Figure 6-9 (a) shows the single and hex routing segments in the design floorplanning and the correspondent segments in the FPGA schematic. Figure 6-9 (b) shows the CLB slice signals and input multiplexor connections in the design floorplanning and the correspondent names in the CLB FPGA schematic. Figure 6-9 (c) shows the output multiplexor signals in the design floorplanning and the corresponding names in the CLB schematic. These figures help one to correlate the signals from the design and the signals in the FPGA schematic, giving the location and placement of both.

6.4 FAULT INJECTION RESULTS

The fault injection was performed in the TMR test design running at 10 and 20 MHz. The report showed that 224 upset bits of 1,663,200 bits in the XQVR300 bitstream had caused an error in the TMR design application execution.

Analyzing the upset bits in the design floorplanning, we observed that a single upset in the routing matrix (GRM) could provoke an undesirable connection between two different signals placed in distinct parts of the FPGA. An example of upset in the GRM that was able to cause an error in the output of the TMR design is located in the major address: 10, frame: 35, frame byte: 46, bit: 5 of the bitstream. Using the equations 3.1 and 3.2, the upset bit in the floorplanning is placed at CLB row 20, column 20 (R20C20) and CLB bit tile: 4. The upset was identified in the IC schematic as I_c_singles.Iw6he1.I377 that means a connection between the single line west 6 and hex line 1, represented in figure 6-10 (a). Apparently, this upset cannot generate an error, because it connects a signal from the comparator of the redundant part 0 to "no" signal. However, the hex line connects the CLB R20C20 to two others CLBs as displayed in figure 6-10 (b) marked by circles. Analyzing the CLB R20C23, for example, we noticed that actually there is a signal connected to this hex line. The signal is from one of the counters of the redundant part 2, as shown in figure 6-10 (c).

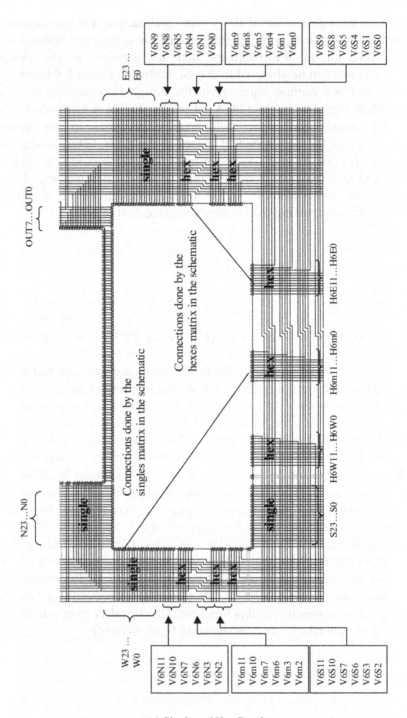

(a) Single and Hex Routing

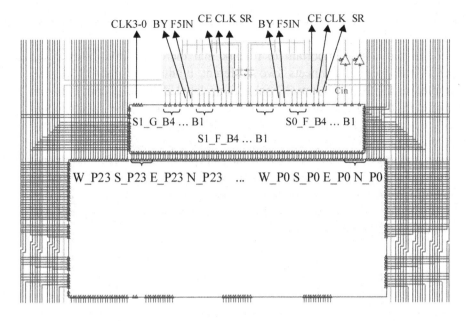

(b) CLB Slices and Input Multiplexors

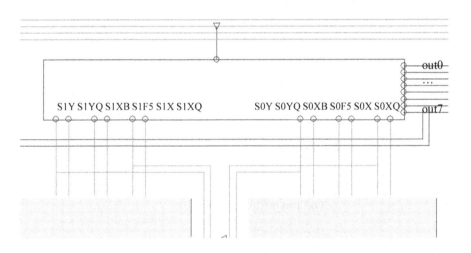

(c) Output Multiplexors

Figure 6-9. CLB Tile representation in the ISE Floorplanning Tool from Xilinx

The analyzed upset bit was characterized by an undesired connection from one bit of the 32-bit counter in a redundant module to a signal from the comparator logic of another redundant module. In this case, both comparators 0 and 2 are going to report an error producing "one" in the error flag. This kind of error would have never occurred if the comparators were placed out of the chip.

In summary, it is important to remark that there is a possibility of an upset if the routing connects two different modules of the TMR. For example, in figure 6-11, upset connections labeled as b, g and f do not interfere in the correct operation of the TMR design. The others could interfere according to the bit that they are affecting because they connect two different logic modules of the TMR. The probability is related to the routing density and logic placement. Dedicated floorplanning for each redundant module of the TMR can reduce the probability of upsets in the routing affecting two or more logic modules. Table 6-5 summarizes the effect of a fault in each FPGA module in the TMR design with no assigned area constraint floorplanning.

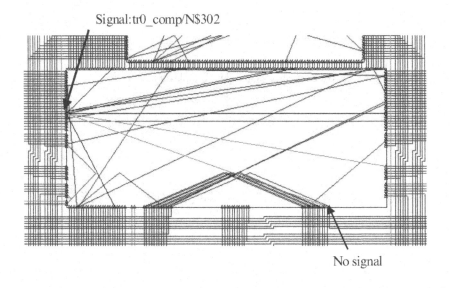

(a) Upset in CLB R20C20

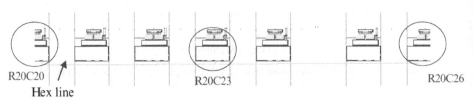

(b) CLBs connected by the hex line

Signal:tr0_comp/N$302
Redundant 0

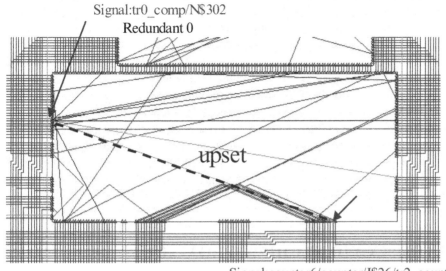

upset

Signal:counter6/counter/I$26/tr2_count(1)
Redundant 2

(c) Undesirable connection detected in CLB R20C20

Figure 6-10. SEU example in the GRM user's design floorplanning

Table 6-5. Upsets Analysis in the Triple Modular Redundancy Approach with No Assigned Area Constraint Floorplanning

Upset Location	Action	Consequences	Upset Correction
LUT	Modification in the Combinational logic	- Error in the redundant part with no error in the TMR design output	By Scrubbing
Routing	Connection or disconnection between any two signals in the design	- Error in the redundant part with no error in the TMR design output - Error in more than one redundant parts with error in the design output	By scrubbing
Customization logic in general	Connection or disconnection between any two	- Error in the redundant part with no error in	By Scrubbing

Upset Location	Action	Consequences	Upset Correction
	signals in the same CLB	the TMR design output - Error in more than one redundant parts with error in the design output	
Flip-flops	Modification in the sequential logic	- Error in the redundant part, no error in the TMR design output	By user correction technique (VHDL)

6.5 THE "GOLDEN" CHIP APPROACH

In order to avoid upset connections between the test design and the comparator test circuitry, a new TMR design based on the "golden" chip approach was implemented in the Virtex® component, where the DUT output signal is compared to the golden design placed outside the chip, figure 6-12. In this case, if a single bit upset in the DUT routing matrix provokes an undesirable connection between two signals from different redundant parts of the design, the TMR will always vote the correct signal to the storage elements and to the output. A bit flip in the customization logic will only be able to generate an error if it upsets the exact same bit in two distinct redundant logic parts, which has an extremely low probability to occur. Moreover, this type of error can be totally avoided with a structured floorplanning of the design placement.

The fault injection experiment using the "golden" chip method was performed in the TMR design running at 25 MHz. The tool has reported "no errors" for all the bits in the bitstream. The radiation characterization results (Lima et al., 2001b; Carmichael; Fuller; Fabula; Lima, 2001) performed at the proton facility in UC Davis show that the Virtex® FPGA has presented the same reliability achieved by the fault injection experiment. However, this was a simple design. More experiments were done to analyze the TMR efficiency in more complex designs, which are presented in the next chapter.

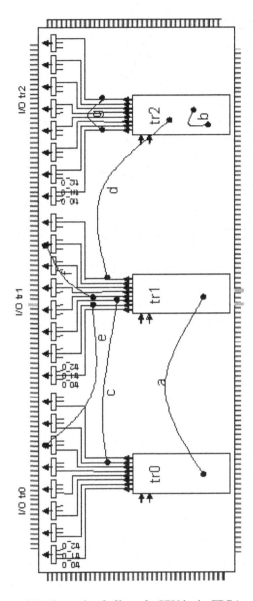

Figure 6-11. Example of effect of a SEU in the FPGA routing

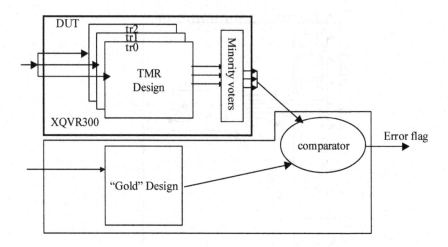

Figure 6-12. "Golden" Chip Method

Chapter 7

DESIGNING AND TESTING A TMR MICRO-CONTROLLER
For SRAM-based FPGA

Micro-controllers implemented in programmable logic platforms are becoming more and more advantageous in order to integrate system-on-a-chip (SOC) improving performance, flexibility and time to market. When a micro-controller is implemented in an SRAM-based FPGA, not only are the registers and memories sensitive to SEU, but also all the programmable logic defined by the FPGA architecture such as the Lookup Tables, routing switches, flip-flops and memories. The previous experiment has shown that TMR can protect designs against SEU in SRAM-based FPGA platforms, for this reason it has been applied to micro-controller architectures too. In addition, a fast time-to-market using commercial off-the-shelf micro-controller architecture for space applications can be achieved by protecting the micro-controller core description and implementing in Virtex® QPRO FPGA.

Following this direction a micro-controller VHDL description developed at UFRGS and presented at (Carro; Pereira; Suzim, 1996; Silva; Lima; Carro; Reis, 1997) was re-used to implement the SEU hardened micro-controller into Virtex® XQVR300 FPGA using the TMR techniques proposed in (Carmichael, 2001). The 8051-like VHDL description is divided into six main blocks as illustrated in figure 7-1. The Finite State Machine (FSM) block implements a counter that generates 24 clock cycles to guide the instruction execution. The Control unit generates all the enable signals for the registers and Arithmetic unit located in the datapath. The Instruction unit generates the microcode word for each instruction. The datapath includes an Arithmetic Logic Unit (ALU) and many registers. There are two 256 byte internal memories, one for the data and the other for the application program.

111

The 8051 micro-controller runs an application based on two 6x6 matrix multiplications at a frequency of 10 MHz. This application performs the multiplication by shifter register and addition. This allows an intensive use of the available memory and internal registers since the operators are read and written many times and both operators and result are stored in the internal memory.

An extra logic circuit was designed to be able to analyze the results of the 8051 after a bit is upset. This block is able to read all the memory data and to send the data to an output pin serially. This output data is compared to the "golden" chip located in a distinct FPGA component. If the data does not match, the comparator circuit sends a flag error to the fault injection tool. Each corrupted bit able to cause an error in the TMR design is reported in a file.

In order to protect the VHDL description against SEU, each logic block has been triplicated, and voters were inserted in all register loops. The datapath, control and instruction logic blocks are mainly throughput logic, and consequently they were just triplicated. The registers in these blocks are constantly being written to avoid being locked in a wrong state. An example of a TMR datapath register with its surround logic in VHDL code is presented in figure 7-2. The vector signals were replaced by an array of 3 vectors (0, 1 and 2) representing the vector signal for each redundant logic part.

All persistent errors (caused by 'weak-keepers') were avoided by using user ground input and user global clock enable. The registered loops located in the state machine and in the counters were protected by TMR with a major voter in each redundant feedback path. All voters were implemented using LUTs. The TMR BRAM component presented previously in figure 3-8 replaced the internal memories. In the DATA memory there is a circuitry able to detect write conflicts in the memory when refreshing. The micro-controller always has the write priority. In the program memory there is no conflict because it is a read only memory from the micro-controller point of view.

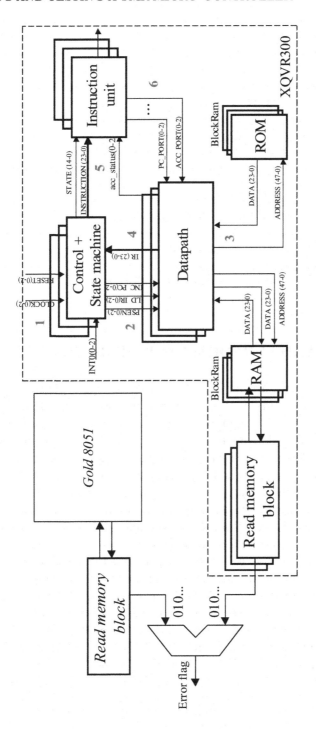

Figure 7-1. TMR 8051 Design Methodology

```
L0: For K in 0 to 2 generate
process   (OP_ACU(K),   reg_alu_out(K),   data_rom(K),   reg_PC_low(K),
data_rd_ram(K))
begin
CASE OP_ACU(K) IS
WHEN "000" => reg_accu_mux(K) <= "00000000";
WHEN "001" => reg_accu_mux(K) <= reg_alu_out(K);
WHEN "010" => reg_accu_mux(K) <= data_rom(K);
WHEN "101" => reg_accu_mux(K) <= reg_PC_low(K);
WHEN others => reg_accu_mux(K) <= data_rd_ram(K);
END CASE; end process;

process (reset_micro(K), clock(K))
begin
if (reset_micro(K)='0') then
    reg_accu(K) <= "00000000";
elsif (clock(K)'event and clock(K)='1') THEN
   if (GCE(K)='1') then
   if (accu_port(K)='1') then
     reg_accu(K) <= reg_accu_mux(K);
   end if; end if;
end if; end process; end generate;
```

Figure 7-2. Example of TMR VHDL code

7.1 AREA AND PERFORMANCE RESULTS

Table 7-1 shows a summary of the TMR design logic overhead in the 8051-like micro-controller. The number of flip-flops in the TMR design has increased 3.6 times. The ratio exceeds 3 because of the extra counters located in the BRAM scrubbing logic. The TMR design contains 3 times the number of BRAM and each one of them has an extra logic of flip-flops and LUTs for voters, counters and logic analyzes. The number of LUTs in the TMR design is approximately 3.6 times bigger than in the standard design. This proportion also exceeds three times because of the voters and the scrubbing logic. Three of the four available global clock buffers in the device are being used for the system clock.

The fault injection experiment was performed at 10 MHz using the test board presented in the previous chapter in figure 6-5. Bit flips were inserted in all 1,663,200 bits of the XQVR300 bitstream. Each fault has remained in the bitstream enough time to run many cycles through the application in the micro-controller. The application is a matrix multiplication. Figure 7-3 shows the values stored in the matrix 1, matrix 2 and the result matrix in the correspondent memory address. The fault injection results showed that less than 1% of the bit upsets could provoke an error in the output of the TMR design, representing a very small cross-section. This 1% is correlated to the faults in the routing that can affect more than one domain in the TMR design. Figure 7-4 shows the floorplanning and routing of the TMR 8051 micro-controller. One can see that the floorplanning occupies a large portion of the entire matrix and the TMR domains are all placed together. The reduction of the number of bits that could provoke an error in the TMR design can be achieved by changing the logic placement. This will be discussed later on the book, as well, the insertion of extra voters to increase the TMR efficiency (Kastensmidt, et al., 2005).

Table 7-1. TMR Logic Overhead in the 8051 (XQVR300)

Item	Standard 8051	TMR 8051
FDCE	127	459
BRAM	2 of 16	6 of 16
TMR BRAM extra logic	-	36 FDCE, 87 LUTs
Inputs	2	12
Outputs	1	3
BUFG	1 of 4	3 of 4
LUTs	757 (12%)	2778 (45%)

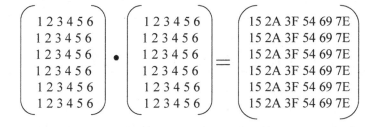

(A) Multiplication of Matrixes

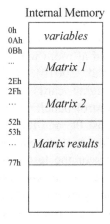

(B) 8051 RAM memory map

Figure 7-3. Application for testing the TMR 8051 micro-controller

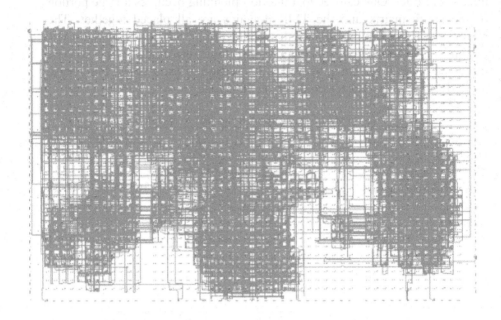

Figure 7-4. TMR 8051 micro-controller routing floorplanning

7.2 TMR 8051 MICRO-CONTROLLER RADIATION GROUND TEST RESULTS

The test was performed at Crocker Nuclear Laboratory at UC Davis, USA. The proton energy and fluxes were measured as incident on the DUT package. All tests were performed at room temperature. More details about the test can be found in (Lima et al., 2001b). The test platform is composed of two AFX V300PQ240-100 daughter cards, a MultiLINX cable used as an interface to a host PC, and a control panel. The system can operate stand-alone or in conjunction with a host PC and test software. The control panel directly communicates with the control chip to specify the mode of operation. Either the control chip or the test software, via the MultiLINX cable, may control the DUT configuration. The control chip also controls the dynamic operation of the DUT and dynamic error detection.

The beam energy was set to 63.3 MeV. The proton flux varied from 8.54E+08 to 1.70E+09 protons/sec-cm^2, in order to ensure a scrubbing rate higher than the error rate. The TMR 8051 design was tested in the dynamic mode and compared to the non-protected design. The tested part was XQVR300 (0.22 µm, 2.5V). The cumulative limit of TID achieved in this test was 116 krads(Si).

The fluence to upset was measured in the design of the no-TMR 8051 and in the TMR version. The first experiment used the test software to readback the bitstream in order to analyze the nature of the dynamic errors. When an error was detected, a readback of the bitstream was initiated and the number of bitstream errors noted alongside the total fluence to functional error. The second experiment compared the design of the TMR 8051 with and without scrubbing. No readback was performed. A logical reset of the flip-flops used in the design would then demonstrate whether the functional error was from configuration/user upsets or the architecture ones. The fluence to upset was measured while the PROM was continually scrubbing the configuration bits.

Experiment 2 tested 3 different approaches in order to demonstrate the benefits of TMR combined with scrubbing, please see figure 7-5. Each test measured the fluence to failure. The no-TMR and TMR designs were tested with and without scrubbing. Table 7-2 presents the TMR 8051 cross-section average for the observed fluence to upset collected in the second experiment.

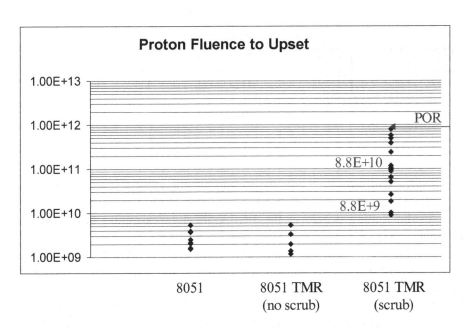

Figure 7-5. Radiation Test Results of the 8051 micro-controller protected by TMR with and without scrubbing

Table 7-2. Virtex® Dynamic Cross-section of TMR 8051

Upset	Hit	Cross-section (cm^2)
Bit	18	6.93E-12
Persistent	2	1.91E-10
POR	0	
Average		2.54E-11

The experiment frequency was set at 9 MHz. The same clock was provided to the scrubbing PROM. The BRAM refreshing performed inside the DUT used the same clock divided by 8. It takes 4 ms to entirely run the two 6x6 matrix multiplications and the internal memory read. The scrubbing takes 22 ms to refresh the whole matrix. And the BRAM refreshing takes 0.2275 ms to refresh all addresses.

In summary, the application runs 4 times during a scrubbing cycle and the BRAM is refreshed 17 times per application cycle as it is illustrated in figure 7-6. The application re-starts with a reset in the micro-controller coming from the read memory logic. In general, bits from the BRAM and the CLB flip-flops (user logic upsets) have the highest refresh rate. The LUTs, customization and routing bits (configuration upsets) are refreshed by the scrubbing rate.

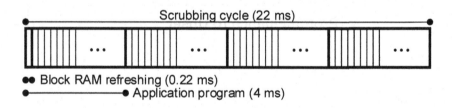

Figure 7-6. Scrubbing and Refreshing Times

An error can only occur in the design functionality if the number of accumulated upsets is enough to overcome the TMR. For example, if an upset in the routing occurs in the first application execution time of the scrubbing cycle, 2 out of 3 legs in the TMR should be able to vote the correct value. However, this undesirable connection or disconnection may affect different parts of the design, generating upsets that can be stored in different redundant parts. All of these upset cells must be refreshed with their original values. If the refreshing rate is such that one cannot avoid the accumulation of upsets, errors are going to be observed in the output. It is important to analyze how the upsets can propagate inside the architecture. In the next application cycle, the CLB flip-flops are going to be reset, however the BRAM are never reset, they have always been refreshed by voting their own values. If the refreshing in the BRAM is not fast enough to avoid accumulation of upsets, a failure can be observed in the output.

The average error rate of TMR 8051 equals to 17 bits/upset (6e-2 upsets/bit/s = 190 upsets/bit/day) and the average scrub rate is 45 per second. This means that in average there are 0.4 upsets per bits scrubbing. This rate could be unsatisfactory, besides the fact that the flux is not always constant (2 or more upsets can occur during a scrubbing cycle) and the upsets can propagate in the architecture generating more upsets. In real applications the scrubbing rate should be at least 2 orders of magnitude higher than the error rate.

In order to improve the results, two solutions can be tried. The first option is to set the proton flux in the radiation facility one order of magnitude or lower, in order to be sure that there is only one upset per scrubbing cycle. In space the flux is much, much lower than the test 99% of the time. However, a very low flux would take a long time to observe each error. The other solution is to speed up the scrubbing frequency. However, the PROM used can only achieve up to 16 MHz with reliable performance.

The TMR technique for SRAM FPGAs was evaluated in Virtex® family using two designs. The first design was a 32-bit counter and the other design was the 8051 micro-controller. The results observed in both designs have proved that the reliability of the TMR is strongly related to the placement of the design in the FPGA matrix. In the first design, the results achieved showed that no errors could be observed in the presence of upsets. The main reason is because the counter design is a simple architecture that does not use embedded memory (BRAM). Consequently, the scrubbing issue is not so evident. In addition, the presented result was based on that specific placement. There is a probability that if another placement had been performed, the results can change.

When a more complex design that uses embedded memory such as the 8051 micro-controller was tested, the probability of upsets in the routing provoking an error in the application results was more eminent, because of its complexity and the scrubbing issue in the embedded memories. Many placements were performed in the TMR 8051, and each one has shown different results in terms of upset bits that could provoke an error in the application results. However, in each case the difference was manly in the routing bit locations that could provoke an error, and not in the number of bits, which was always around 1% of the bits of the bitstream.

Based on the analyzed results, the TMR technique in SRAM based FPGAs can substantially improve the reliability of the design, but there is a low probability of error caused by upsets in the routing. This result is correlated with the logic placement. The solution can be obtained by using a fault injection tool combined with a dedicated placement, routing and number of voters, in order to achieve 100% reliability.

In (Kastensmidt, et al., 2005), the number of voters that should be used in a TMR design is discussed. The majority voters perform a very important task in the TMR approach, because they are able to block the effects of an upset through the logic at the final output. In this way, the voters can be placed in the end of the combinational and sequential logic blocks, creating barriers for the upset effects. The problem is to determine the optimal partition of the TMR logic that must be voted inside the circuit in order to reduce the probability that upsets in the routing affect two distinct redundant parts that are voted by the same voters. A small size block partition requires a large number of majority voters that may be too costly in terms of area and performance. On the other hand, placing only voters at the last output increases the probability of an upset in the routing affecting two distinct redundant logic parts overcoming the TMR.

If the redundant logic parts tr0, tr1 and tr2 (represented in figure 7-7 after the TMR register with voters and refresh) are partitioned in smaller logic blocks with voters, a connection between signals from distinct redundant parts could be voted by different voters. Notice that now the upset "b" can not provoke an error in the TMR output, which increases the robustness of the TMR in the presence of routing upsets without being of concern to floorplanning. The problem is to evaluate the best size of the logic to achieve the best robustness. Fault injection tool can help to identify the best amount of TMR voters in the final design.

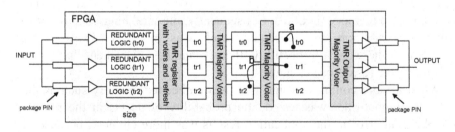

Figure 7-7. Triple Modular Redundancy (TMR) Scheme with Logic Partition in the FPGA

Although the insertion of extra voters, which provokes a partition in the logic, can improve the TMR reliability, it does not avoid completely the chance of a upset in the routing affect two or more TMR domains. In (Bernardi, et al., 2004), it is presented a specific reliability-oriented place and route algorithm called (RoRA), which enforces layout-level design constraints and guarantees that one SEU affects only one module domain in the replicated TMR system. In general, RoRA starts by reading a technology-mapped circuit and by dividing each logic functions in different sets (i.e. in the case of the TMR, in three sets S0, S1 and S2). Then it performs the place and route operations implementing two fault-tolerant rules in order to avoid multiple errors that can involve two sets Si and Sj where i ≠ j:

1. All the logic functions of a set must be placed in distinct CLBs. In such a way that each CLB contains logic functions belonging to one set only.

2. The interconnections between the logic functions must be routed in such a way that, given the corresponding routing architecture, each new PIP that is added/deleted to/from the routing architecture cannot provoke shorts between different logic functions sets or open in different connections belonging to different logic function sets.

Results have presented that a combination of TMR, RoRA and additional voters may ensure high levels of reliability in many case studied circuits (Kastensmidt, et al., 2005). In addition, a research has been started on investigating the contribution of routing duplication to improve reliability in TMR designs in SRAM-based FPGA, specially the usage of routing duplication on the majority voter parts. Preliminary results about routing duplication are presented in (Kinzel, et al., 2005).

Chapter 8

REDUCING TMR OVERHEADS: PART I
By Combining Hardware and Time Redundancy

The TMR technique is a suitable solution for FPGAs because it provides a full hardware redundancy, including the user's combinational and sequential logic, the routing, and the I/O pads. However, it comes with some penalties because of its full hardware redundancy, such as area, I/O pad limitations and power dissipation. Although these overheads and limitations could be reduced by using some architectural SEU mitigation solutions such as hardened memory cells, EDAC techniques and standard TMR with single voter, as presented in chapter 4, these solutions are very costly, because they require modifications to the matrix architecture of the FPGA (NRE cost). Many applications can accept the limitations of the TMR approach but some cannot. The main limitations are:

- The number of I/O pads available for designers is reduced by three times, because each input and output of each TMR redundant block (tr0, tr1, tr2) should have its own input and output pads. The number of dedicated clock resource segments for the routing available is also reduced by three times, because each input and output of each TMR redundant block (tr0, tr1, tr2) should have its own clock.

- The size of the combinational logic in the design is multiplied by three times, and this also happens in the sequential logic, where each storage cell must be replaced by three storage cells, with three voters and multiplexors.

- The embedded memory also needs to be triplicated and refreshed using extra logic.

- There is a delay overhead inserted by the voters.

-       The power consumption is increased by three times, as all input and output pins as well as the combinational and sequential logic are triplicated.

A new high-level fault-tolerant technique is introduced in this chapter that combines time and hardware redundancy, with some extra features able to cope with SEU in SRAM-based FPGAs. This technique allows the reduction of the number of I/O pads, and consequently power dissipation in the interface. The main idea is to reduce the hardware overhead, which in the case of the TMR is three times more to some point close to twice the original area, maintaining the same reliability.

The possibility of applying time redundancy combined with hardware redundancy for FPGAs looks interesting to reduce the costs of using full hardware redundancy (TMR) and to improve reliability (less sensitive area). Potentially, the use of duplication with comparison (DWC) combined with time redundancy may reduce area and pin count and consequently power dissipation in the I/O pads, the main drawbacks of the TMR approach. But there are two problems to be solved. First, previous techniques based on time redundancy can only be used to detect upsets, and not upsets that become permanent, as is the case of SRAM based FPGAs. Second, in the FPGA, usually DWC can only detect an upset, but in this case, it is not only sufficient to detect an upset, but one also must be able to vote the correct value in order to ensure the correct output. In the next section, we present a technique based on time and hardware redundancy for SRAM-based FPGA that takes into consideration the above problems to reduce pin count, area and power dissipation.

8.1 DUPLICATION WITH COMPARISON COMBINED WITH TIME REDUNDANCY

Time redundancy by itself can only detect transient faults (Nicolaidis, 1999; Anghel; Alexandrescu; Nicolaidis, 2000). The same occurs with duplication with comparison (DWC), which can also only detect faults. However, the combination of time redundancy and DWC can provide an interesting upset evaluation, which can not only detect the presence of a fault, but also recognize in which redundant block the upset has occurred. Figure 8-1 shows the detailed scheme. There are two redundant blocks: dr0 and dr1. In this way, upsets in the combinational logic can be detected and voted before being latched.

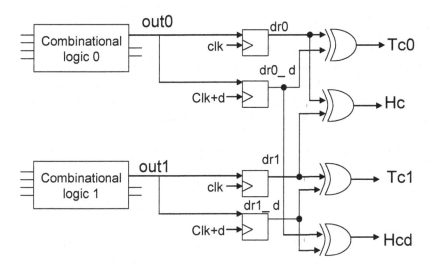

Figure 8-1. Time and Hardware Redundancy Schematic for Upset Detection

Four values are stored in the auxiliary latches (dr0, dr0_d, dr1 and dr1_d), two from each redundant block collected at different instants. Two latches store the dr0 and the dr1 outputs at the clock edge and two latches store the dr0 and dr1 outputs at the clock edge plus a delay d. As a consequence, there are four outputs of comparators in the scheme: Tc0 is the time redundancy comparator from redundant block 0, Hc is the hardware redundancy comparator at the clock edge, Tc1 is the time redundancy comparator from redundant block 1 and Hcd is the hardware redundancy comparator at the clock+d edge. Analyzing the sixteen possibilities of output combinations of dr0, dr0_d, dr1 and dr1_d, eight different syndromes are recognized, as presented in table 8-1. Analyzing the syndromes from the table, it is possible to see the temporal effect of an upset in the FPGA. The steps are basically no fault, upset effect in redundant block 0 (dr0) or block 1 (dr1), permanent effect in redundant block 0 (dr0) or block 1 (dr1), recovery upset effect in redundant block 0 (dr0) or block 1 (dr1), and no fault.

An upset in redundant block 0, syndrome 1001, is characterized by a transient variation in the output (Tc0 = 1) with no changes at output dr1 (Tc1 = 0), and in addition Hc = 0 and Hcd = 1. An upset occurrence in dr1 is recognized in an equivalent way, where Tc1 = 1 and Tc0 = 0. There are many other syndromes that are not commonly seen in an ASIC environment, only in FPGAs. One example is the permanent effect of an upset, syndrome 0101. By analyzing this syndrome, it is not possible to conclude which redundant block has the correct value and which does not. The previous syndrome characterized by the transient effect detection is necessary to vote the correct path. This phenomenon characterizes the necessity of a state machine to vote the correct value. This technique considers only one upset per design at once, either in redundant block 0 or in redundant block 1. An implementation with an assigned area constraint may avoid the occurrence of a fault in the redundant block 0 at the same time as a fault in redundant block 1 (syndrome 1010). The identification of this syndrome can be used as a flag to show that upsets have overcome the DMR scheme.

Table 8-1. Syndrome Analysis in the Double Modular Redundancy Approach

dr0	dr0_d	dr1	dr1_d	Tc0	Hc	Tc1	Hcd	Syndrome
0	0	0	0	0	0	0	0	No fault
0	0	0	1	0	0	1	1	Fault dr1 (stage 1, transient)
0	0	1	0	0	1	1	0	Fault recovery dr1
0	0	1	1	0	1	0	1	Fault dr0 or dr1 (stage 2, permanent)
0	1	0	0	1	0	0	1	Fault dr0 (stage 1, transient)
0	1	0	1	1	0	1	0	Fault dr0 and dr1 (stage 1, transient)
0	1	1	0	1	1	1	1	Fault dr0 or dr1, recovery dr0 or dr1
0	1	1	1	1	1	0	0	Fault recovery dr0
1	0	0	0	1	1	0	0	Fault recovery dr0
1	0	0	1	1	1	1	1	Fault dr0 or dr1, recovery dr0 or dr1
1	0	1	0	1	0	1	0	Fault dr0 and dr1 (stage 1, transient)
1	0	1	1	1	0	0	1	Fault dr0 (stage1, transient)
1	1	0	0	0	1	0	1	Fault dr0 or dr1 (stage 2, permanent)
1	1	0	1	0	1	1	0	Fault recovery dr1
1	1	1	0	0	0	1	1	Fault dr1 (stage1, transient)
1	1	1	1	0	0	0	0	No fault

The DWC with time redundancy proposed technique for the combinational blocks, illustrated in figure 8-2, combines duplication with comparison and time concurrent error detection to identify combinational upsets in FPGAs. DWC with time redundancy tolerates upsets without system interruptions. The combinational logic is duplicated and there is a voter circuit able to detect an upset and to identify which redundant block should be connected to the CLB flip-flops. The upsets in the combinational logic are corrected by scrubbing, while upsets in the CLB flip-flops are corrected by the TMR scheme. It is important to notice that for upset correction, scrubbing is continuously performed, to ensure that only one upset has occurred between two reconfigurations in the design.

When the circuit is reset, the state machine starts in state 0 and it is persistently monitoring the redundant block 0, while the redundant block 1 is the spare. At this point, an upset in the redundant block 1 will not affect the system, and it will be corrected by the periodic scrubbing. If an upset occurs in the redundant block 0, the state machine recognizes the fault, and the operation switches to the spare path, the redundant block 1. The upset in the redundant block 0 will be soon corrected by the scrubbing, while now the system is operating with the redundant block 1. At this point, the state machine is constantly monitoring the redundant block 1, looking for upsets and the redundant block 0 is the spare. Upsets in the redundant block 0 will be corrected by scrubbing.

As the fault detection technique used for this method is able to identify only transient faults, it is necessary to have an observation period to detect the fault occurrence and consequently the faulty module (dr0 or dr1). The size of the observation period of a fault occurrence is referred to as the clock delay d. As a result, the transient fault observability occurs between clock and clock+d. Outside this observation period, the fault is seen as permanent and the faulty module can not be recognized, only the presence of a fault can be detected. The percentage of faulty module detection is related to d. As the observation period (d) becomes greater, the probability of faulty module detection becomes higher. One can use d as half of one clock cycle. The performance penalty of this method is related to the time duration of the fault observability (d).

The registers from the sequential logic store the combinational logic outputs at clock plus d, plus the delay from the upset detection circuit, totaling a new delay d'. The latches from the concurrent upset detection state machine will also store the next state at clock+d'. In order to simplify the number of clocks in the design, one possibility is to reduce the frequency of the design by two. In this way, the combinational output is stable at the clock falling edge. At this time, the value is captured for the future comparison. The fault observability period is until the next clock rising edge where the correct redundant logic is voted. Figure 8-3 illustrates two fault effects, one occurring during the propagation time and one occurring during the observation time.

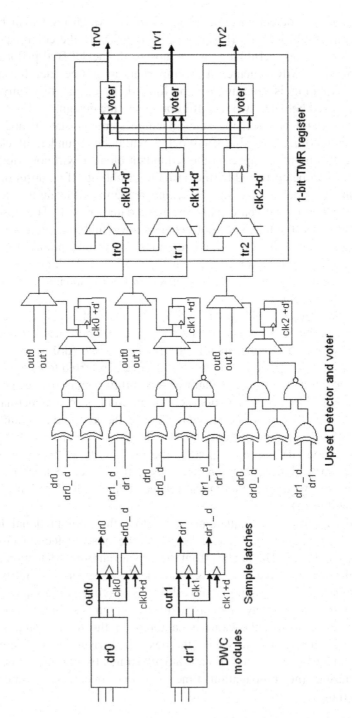

Figure 8-2. DWC with time redundancy proposed technique scheme for one bit output

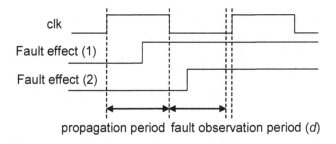

Figure 8-3. Fault effect in the clock period

If a fault effect occurs during the propagation period, the DWC with time redundancy scheme will detect an error but will not be able to recognize which redundant block (dr0 or dr1) is faulty. However, some fault effects occurring during the propagation period can be tolerated, if they affect the spare redundant logic that is not being observed at that time. The fault can be corrected in the next scrubbing and no error may occur. If a fault effect occurs during the observation time, the DWC with time redundancy scheme will be able to detect the output variation and vote the fault-free redundant module (dr0 or dr1). Faults in the observation time will always be correctly voted, except for those whose effect will not be manifested at the time, for instance, a fault stuck at one in a node that already has the logic value one because of the input vectors.

Some constraints must be observed for the perfect functioning of the technique. The constraints are the same as TMR:

-   There must be only one upset per dual modular redundancy (DMR) combinational logic, including the state machine detection and voting circuit, consequently it is important to use some assigned area constraints to reduce the probability of short circuits between redundant block 0 and 1 (dr0 and dr1).

-   The scrubbing rate should be fast enough to avoid accumulation of upsets in two different redundant blocks.

-   Upsets in the detection and voting circuit do not interfere with the correct execution of the system, because the logic is already triplicated. In addition, upsets in the latches of this logic are not critical, as they are refreshed in each clock cycle. Assuming a single upset per chip between scrubbing, if an upset alters the correct voting, it does not matter as long as there is no upset in both redundant blocks.

This technique can be used as an additional option for the TMR technique for designing reliable circuits in FPGAs with pads and power reduction. Because the combinational circuit is just duplicated, inputs and outputs can be duplicated instead of triplicated, as in the TMR approach. However, it is important to notice that the TMR inputs related to the user's sequential logic used in the CLB flip-flops are not changed, still requiring triple input clocks, reset and user vdd and gnd (Carmichael, 2001).

The upset detector and voter circuit can be optimized in terms of area. In figure 8-2, the upset detector and voter circuit are represented for only one bit. However, it is possible to use the circuit for groups of bits. In this way, only one state-machine per TMR redundant part for each group of bits is necessary, as presented in figure 8-4. Another possible optimization is to use a single state machine to vote just the input of the redundant block 2 of the TMR register, as presented in figure 8-5. In this way, a fault in one of the combinational redundant blocks (dr0 or dr1) is voted to the tr2 input, assuring the correct operation. A fault in this upset detection and voter block will corrupt just the redundant block 2 of the TMR (tr2), consequently, tr0 and tr1 will still vote the correct value. The presented scheme also shows the clock optimizations, where the sample storage cells are latched at the clock falling edge (clk0, clk1) and the state machine of the upset detection block is latched at the clock rising edge (clk2). The three clocks are the same, and are all connected outside the FPGA chip.

In summary, the final DWC with time redundancy scheme is composed of:

-    Two redundant blocks of the combinational logic.

-    A set of sample latches related to the number of output bits of each redundant block, which is used to capture the value at the clock falling edge.

-    Upset detection block, which is continuously monitoring a variation between the captured value and the combinational output during the observation period (clock low level).

-    The corrected redundant part is voted just before the next clock rising edge, where the TMR redundant part 2 from the register stores the fault-free redundant logic (dr0 or dr1).

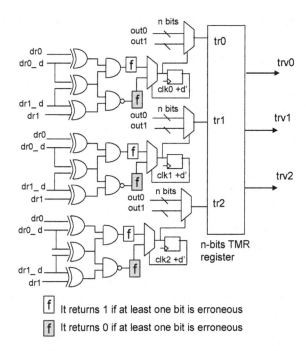

Figure 8-4. Upset detector and voter circuit area optimization using group of n bits

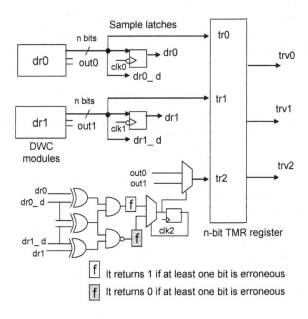

Figure 8-5. Upset detector and voter circuit area optimization using a single state machine for a group of n bits

8.2 FAULT INJECTION IN THE VHDL DESCRIPTION

The DWC with time redundancy scheme was validated by fault injection methodology in a prototype board using VHDL. The fault injection system described in VHDL was specifically developed to test the proposed technique (Delong; Johnson; Profeta, 1996; Lima et al., 2001a). Results were emulated in an AFX-PQ249-110 board using a XCV300 part. Some area comparisons between the proposed approach and TMR were also performed using Xilinx implementation tools. We use multipliers as combinational circuit case studies, and FIR canonical filters as sequential circuit case studies.

Fault injection in VHDL combined with the full emulation in a FPGA platform was used to characterize and validate the technique. The proposed fault injection system emulates single event upsets in memory related components (single flip-flops or latches, registers and memories) designed in high-level description and in the combinational logic nodes. The whole system is a run-time fault injection mechanism that is performed during the prototype execution without interrupting the design application. It injects one fault per execution. This approach does not concern the mean time between failures (MTBF), in other words, we are not considering more than one fault per execution time or the fault occurrence frequency. The approach aims to emulate a single upset per execution, and to validate the efficiency of SEU mitigation techniques. However, this technique is completely customizable, and it can inject as many upsets as wanted per execution.

The developed fault injection system is divided into 3 main design blocks, figure 8-7:

-   Fault injection Control block: generates all the fault enable signals to all register, memories and combinational nodes. It also chooses the time and location of the injected transient bit flip fault or a stuck at fault (in the case of FPGA),

-   Device Under Test (DUT) core: the modified design core. Fault injection paths are added to the design in order to inject bit flips or stuck at one in all SEU sensitive parts and logic nodes,

-   Monitor block: responsible for monitoring the results of the DUT core in order to analyze the effects of each inserted fault.

Figure 8-8 shows some schemes for the fault injection in the memory, register and combinational nodes. The main advantages of this approach compared to a software based method are its high flexibility of fault injection parameters (time, location and fault value), fast turnaround time and free access to all sensitive parts of the design.

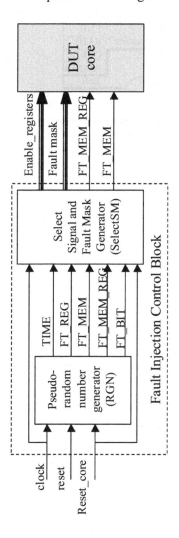

Figure 8-7. Schematic of the fault injection generator block

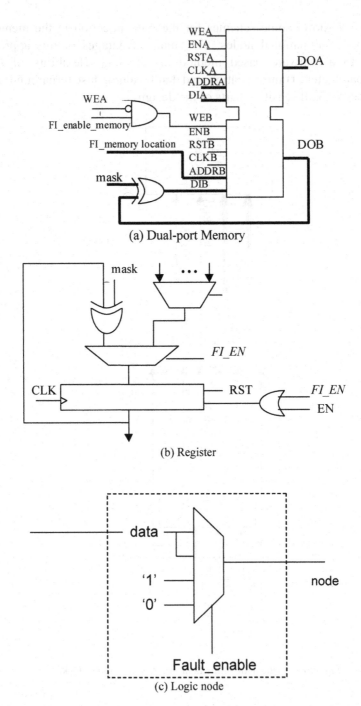

(a) Dual-port Memory

(b) Register

(c) Logic node

Figure 8-8. Example of the mechanisms used to inject faults in the design

The fault injection system is able to randomly choose the fault time, the fault node and the redundant block. In order to test the duplication method in the combinational logic, stuck at one and stuck at zero faults were injected in all nodes, emulating the bit-flip in a SRAM cell in the FPGA architecture (permanent effect of a SEU). There is a reset fault signal that works as a scrubbing, cleaning up the fault. A 2x2 bit multiplier with and without a register at the output was chosen for this first evaluation. It is possible to inject a set of faults in all the user's combinational nodes of the example circuit, covering several time intervals in the clock cycle, and to emulate the scrubbing between faults. The multiplier input vectors were also randomly generated.

Figure 8-9 exemplifies a fault stuck at one that was inserted in the redundant block 0 during the observation time. There is one point of data acquisition at the clock falling edge, just after the combinational output has stabilized. The fault must be detected before the next clock rising edge (clock+d). The fault effect between these two points can be easily detected, and the correct redundant block can be voted. However, upset effects located extremely near the clock rising edge of the register, or during the propagation time cannot be voted, but they can be detected. This limitation of the detection of a fault is due to the impossibility of distinguishing a data disparity coming from a fault or from the input variations in the redundant block 0 and redundant block 1. As the effect of an upset in the user's combinational logic in an FPGA is permanent, all the results from the redundant block 0 after the fault effect are erroneous until the next scrubbing takes place.

Fault injection results show the reliability of the presented method. There were 128 stuck at one and 128 stuck at zero faults inserted in a random single node (ranging from 0 to 7) at a random instant of the clock cycle in a 2x2 bit multiplier that could occur during the propagation or the observation time. Among the stuck at one faults, 113 of them were detected and tolerated, either because they were correctly voted or because the fault did not affect the correct design output. Among the stuck at zero faults, 121 of them were also detected and tolerated, either because they were correctly voted or because the fault did not affect the correct design output.

The injected faults during the observation time that generated an error were the ones where the effect could not be observed by the input vectors at that time. Faults occurring during the propagation time were detected and some of them were also tolerated. The tolerated faults are the ones that occurred in the spare redundant block. When the upset effect happens during the propagation time, the proposed scheme is not capable of detecting in which redundant block the fault has occurred, only detecting that the system is in error. Consequently, after fault detection with no correction (syndrome

0101), the system should be reinitialized or some results should not be considered.

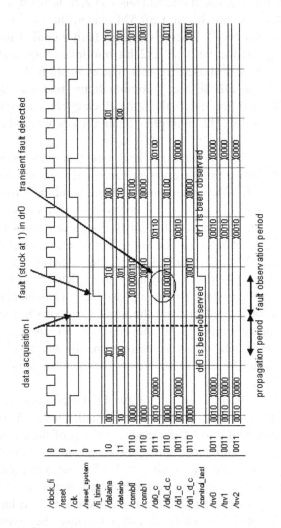

Figure 8-9. Simulation Analysis of a fault injection in the DMR with time redundancy scheme implemented in a 2x2 bits multiplier

8.3 AREA AND PERFORMANCE RESULTS

Table 8-2 presents area results of 2x2, 8x8 and 16x16 bit multipliers, implemented in the XCV300 FPGA using no tolerance technique, TMR technique and DWC with time redundancy in order to reduce pin count. All of the multipliers were synthesized with a register at the output. Results show that it is possible not only to reduce the number of I/O pins but also the area, according to the size of the combinational logic block. Note that the 16x16 bit multiplier protected by TMR could not be synthesized in the prototype board that uses a Virtex® part with 240 I/O pins (166 available for the user). However, the same multiplier implemented by the proposed technique could fit in the chip, also occupying less area.

Table 8-2. Example of combinational circuit: Multiplier Implemented in XCV300-PQ240 FPGA

Multipliers	Standard			TMR			DWC with time redundancy		
	2x2	8x8	16x16	2x2	8x8	16x16*	2x2	8x8	16x16
Combinational Input Pins	4	16	32	12	48	96	8	32	64
Sequential Input Pins	2	2	2	12	12	12	12	12	12
Output Pins	4	16	32	12	48	96	12	48	96
Number of 4-input LUTs	4	156	705	16	514	2002	33	440	1504
Number of ffs	4	16	32	12	48	96	21	81	161

I/O pins were out of range for the TMR approach, the part XCV300-BG432 was used.

There is a constant area in this proposed method, resulting from the upset detection and voter block. Consequently, the proposed approach will only show a smaller area than TMR when the area of the combinational logic related to the third redundant part of the TMR that is suppressed is larger than this constant cost. However, this technique can be used in I/O circuitry, to ensure pin count reduction in critical pin count designs.

A canonical FIR filter circuit was chosen as a sequential case study circuit for the proposed technique. Digital filters such as the finite-length impulse response (FIR) filter are typically used in many DSP-based systems applications that usually use FPGAs, such as image and voice-processing applications. Figure 8-10 shows the scheme of a canonical filter of 5 taps. The multipliers were designed with constant coefficients, resulting in an optimized area. The registers are protected by TMR, figure 8-11, while the combinational logic (multipliers and adders) is protected by DWC with time redundancy technique. The upset detection and voter block is placed at the outputs, and it votes the correct pad output from dr0 or dr1, as shown in figure 8-12.

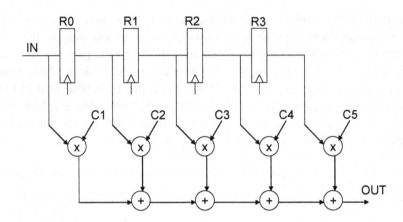

Figure 8-10. Example of FIR Canonical Filter of 5 taps scheme

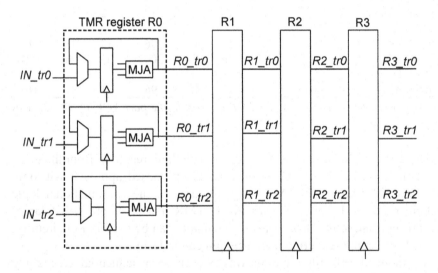

Figure 8-11. Filter registers protected by TMR

An 8-bit FIR canonical filter of 9 taps was synthesized in an XCV300 FPGA to evaluate area and pin count. The multiplier coefficients are: 2, 6, 17, 32 and 38. Table 8-3 presents area results of this filter using no tolerance technique, TMR technique and the proposed technique. Results show that the 9 taps FIR canonical filter occupies 22% less area in the FPGA if protected by DWC and time redundancy instead of by TMR. The results also present a reduction of 20% in the pin count compared to TMR.

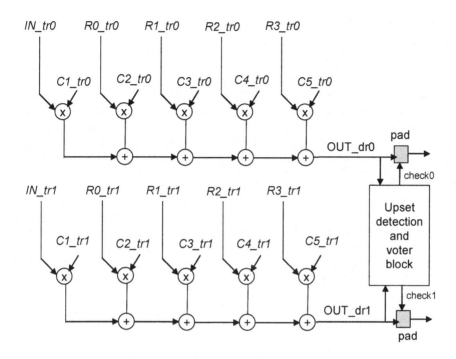

Figure 8-12. Filter adders and multipliers protected by DWC with time redundancy

Table 8-3. Example of Sequential circuit: FIR canonical filter of 9 taps implemented in XCV300-PQ240 FPGAs

	Standard	TMR	DWC with time redundancy
Combinational Input Pins	8	24	24
Sequential Input Pins	3	15	15
Output Pins	16	48	32
Number of 4-input LUTs	265	948	741
Number of ffs	64	192	225

According to the user's application requirements, the designer will be able to choose between a full hardware redundancy implementation (TMR) or a mixed solution where duplication with comparison is combined to concurrent error detection to reduce pins and power dissipation in the interface, as well as area, as shown in previous examples. Figure 8-13 shows some implementations combining TMR and DWC with time redundancy. It is possible to use this new technique only in the interface of the FPGA, in this way reducing pins, as shown in figure 8-13(a). DWC with time redundancy can also be used along the design as presented in figure 8-13(b) to reduce the number of I/O pads and also area for large combinational circuits, as presented in table 8-2 and table 8-3.

Sequential circuits such as counters and state machines are more suitable to be protected by TMR, as the combinational logic is small compared to the sequential logic. The proposed technique is an alternate method to protect combinational circuits, as it is necessary to insert a concurrent error detection block. On the other side, large combinational logic blocks can be easily found in many applications. For example, microprocessors are composed of combinational logic such as the Arithmetic and Logic Unit, multipliers and the micro-instruction decoder.

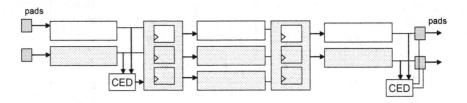

(a) DMR with time redundancy implementation in the interfaces

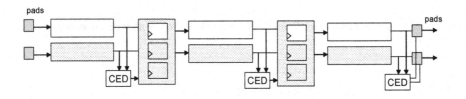

(b) DMR with time redundancy implemented in the entire circuit
Figure 8-13. Evaluation schemes of the TMR and the DWC with time redundancy approach

The proposed approach was validated by fault injection in a Virtex® prototype board using VHDL. Upsets were randomly inserted in the user's combinational logic nodes to emulate faults in the logic. The fault injection procedure was developed in VHDL, and it represents the effect of a SEU in a SRAM-based FPGA, where it has a transient effect followed by a permanent effect. Experiments in a 2x2 bit multiplier showed that 100% of the faults can be detected and 234 of the 256 injected stuck at zero and stuck at one faults (91%) were tolerated, either because they were correctly voted before being captured by a CLB flip-flop or that specific faults did not affect the correct design output.

Although the time redundancy technique can be successfully used to reduce pin count and area overhead over a full hardware redundancy, the transient concurrent error detection technique is not able to correct 100% of the faults occurring in FPGAs. Another penalty of this method is performance overhead because of the observation time. The evolution of this work investigates the use of modified time redundancy technique based on permanent fault detection to improve fault correction and to reduce the performance penalty at each clock cycle, as it is shown in the next chapter.

Chapter 9

REDUCING TMR OVERHEADS: PART II
By Combining Concurrent Error Detection with Duplication with comparison tecniques

Time redundancy by itself cannot detect 100% of the faults in an SRAM-based FPGA because of the permanent effect of the faults. Consequently, it is necessary to continue investigating a technique able to detect the presence of permanent faults in the logic circuit. (Lubaszewski; Courtois, 1998) discusses the reliability and the safety of TMR scheme compared to self-checking-based fault-tolerant schemes. The reported experimental results show that the higher the complexity of the module, the greater the difference in reliability between self-checking and TMR. In summary, the self-checking fault-tolerant scheme can achieve a higher reliability in comparison to the TMR if the self-checking overhead bound of 73% is not exceeded.

The idea of using self-checking fault-tolerant scheme can be extended for FPGAs by using the duplication with comparison (DWC) method combined with concurrent error detection (CED) technique. Figure 9-1 presents the scheme, called hot backup DWC-CED. The CED is able to detect which module is faulty in the presence of an upset, and consequently, there is always a correct value in the output of the scheme, because the mechanism is able to select the correct output out of two.

In the case of SEU detection in SRAM-based FPGAs, the CED must be able to identify permanent faults in the redundant modules. The CED works by finding the property of the analyzed logic block that can help to identify an error in the output in the presence of a permanent fault. There are many methods to implement logic to detect permanent faults, most solutions are based on time or hardware redundancy and they manifest a property of the logic block that is being analyzed.

143

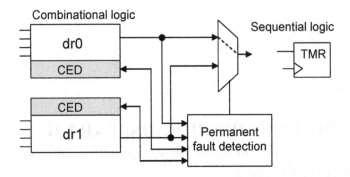

Figure 9-1. DWC combined with CED scheme

The CED scheme based on time redundancy recomputes the input operands in two different ways to detect permanent faults. During the first computation at time t0, the operands are used directly in the combinational block and the result is stored for further comparison. During the second computation at time t0+d, the operands are modified, prior to use, in such a way that errors resulting from permanent faults in the combinational logic are different in the first calculation than in the second and can be detected when results are compared. These modifications are seen as encode and decode processes and they depend on the characteristics of the logic block. The scheme is presented in figure 9-2.

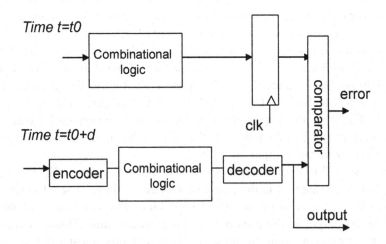

Figure 9-2. Time redundancy for permanent fault detection

If an output mismatch occurs, the output register will hold its original value for one extra clock cycle, while the CED block detects the permanent fault. After this, the output will receive the data from the fault free module until the next reconfiguration (fault correction). The important characteristic of this method is that it does not incur a high performance penalty when the system is operating free of faults or with a single fault. The method just needs one clock cycle in hold operation to detect the faulty module, and after that it will operate normally again without performance penalties. The final clock period is the original clock period plus the propagation delay of the output comparator. Sample registers are latched at the rising clock edge and the user's TMR registers are latched at the rising clock+d edge.

Many techniques to encode and decode were proposed in the literature to detect permanent faults (Johnston; Aylor; Hana, 1988; Patel; Fung, 1996; Avizienis, 1971), some based on time redundancy, such as bit-wise inversion, re-computing with shift operands (RESO) and re-computing with swapped operands (REWSO); and some based on hardware redundancy, such as parity prediction and module code.

9.1 DWC-CED TECHNIQUE IN ARITHMETIC-BASED CIRCUITS

The combination of DWC technique and CED blocks enabling one to detect permanent faults provides a new high-level SEU mitigation technique for FPGAs. Two clock cycles are needed to identify a permanent fault in the combinational logic module. However, this extra time does not occur at every clock operation in our approach. Using DWC combined with CED for permanent faults, it is possible to take advantage of the simple comparison at the output of the duplication scheme to inform whether it is necessary to re-compute the data for permanent fault detection. The re-computation is needed only when a mismatch of the outputs occurs. This method has been named duplication with comparison combined to concurrent error detection block (DWC-CED).

Figure 9-3 shows the multiplier scheme proposed for an arithmetic module, in the present case study. There are two multiplier modules: mult_dr0 and mult_dr1. There are multiplexors at the output able to provide normal or shifted operands. The output computed from the normal operands is always stored in a sample register, one for each module. Each output goes directly to the input of the user's TMR register. Module dr0 connects to register tr0 and module dr1 connects to register tr1. Register tr2 will receive the module that does not have any fault. By default, the circuit starts passing the module dr0. A comparator at the output of register dr0 and dr1 indicates an output mismatch (Hc). If Hc = 0, no error is found and the circuit will continue to operate normally. If Hc = 1, an error is characterized and the operands need to be re-computed using the RESO method to detect which module has the permanent fault. The detection takes one clock cycle.

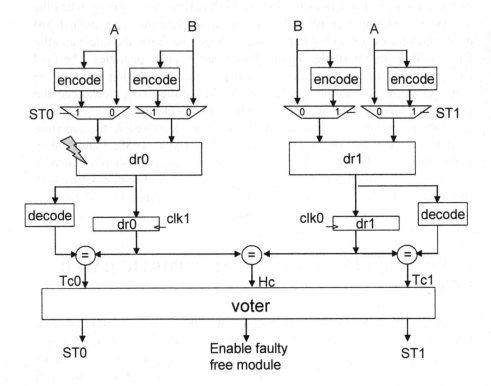

Figure 9-3. Fault tolerant technique based on DWC combined with CED for SRAM-based FPGAs

In the case of a registered output, each output goes directly to the input of the user's TMR register. Figure 9-4(a) illustrates the logic scheme. Module dr0 connects to register tr0 and module dr1 connects to register tr1. While the circuit performs the detection, the user's TMR register holds its previous value. While the circuit performs the detection, the TMR register holds its previous value. When the faulty free module is found, register tr2 receives the output of this module and it will continue to receive this output until the next chip reconfiguration (fault correction). By default, the circuit starts passing the module dr0. In the case of a non-registered output, the signals can be driven directly to the next combinational module or to the I/O pads, as shown in figure 9-4(b).

Let's consider two different fault situations when the output is saved in a TMR register. In one, the fault occurs in module dr0 (Mult_dr0). Hc indicates that there is an output mismatch; Tc0 indicates that module dr0 is faulty and Tc1 indicates that dr1 is fault free. This analysis takes one clock cycle. Consequently, the permanent fault detection block selects dr1 for the tr2 input. Note that the value stored in the user's TMR register is held for one cycle while the scheme identifies the faulty free module. In the second case, a fault occurs in the module dr1 (Mult_dr1), similar to the previous example, Hc indicates that there is an output mismatch; Tc0 indicates that module dr0 is fault free and Tc1 indicates that dr1 is faulty. The permanent fault detection block selects dr0 for the tr2 input.

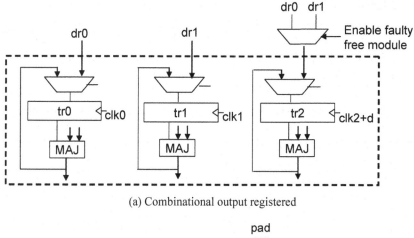

(a) Combinational output registered

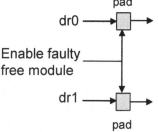

(b) Combinational output in the pad

Figure 9-4. Examples of implementations with the combinational output registered and in the pads

Note that in both methods, TMR and the proposed technique, the upsets in the user's combinational logic are corrected by scrubbing, while upsets in the user's sequential logic are corrected by the TMR scheme used in the CLB flip-flops. It is important to notice that for upset correction the scrubbing is continuously performed, to guarantee that only one upset has occurred between two reconfigurations in the design. Some constraints must be observed for the perfect functioning of the technique, same as TMR: there must not be upsets in more than one redundant module, including the state machine detection and voting circuit, consequently it is important to use some assigned area constraints to reduce the probability of short circuits between redundant module dr0 and dr1. The scrubbing rate should be fast enough to avoid accumulation of upsets in two different redundant blocks. Upsets in the detection and voting circuit do not interfere with the correct execution of the system, because the logic is already triplicated. In addition, upsets in the latches of this logic are not critical, as they are refreshed in each clock cycle. Assuming a single upset per chip between scrubbing, if an upset alters the correct voting, it does not matter, as long as there is no upset in both redundant blocks.

In the proposed method, the area reduced by the design compared to the TMR is the area of one user's combinational logic module and the number of inputs that is reduced from 3 times to 2 times the original number. This technique can be used as an additional option for the TMR technique for designing reliable circuits in FPGAs with pads and power reduction. Because the combinational circuit is just duplicated, inputs and outputs can be duplicated instead of triplicated, as in the TMR approach. However, it is important to notice that the TMR inputs related to the user's sequential logic used in the CLB flip-flops are not changed as triple clocks, reset, etc.

In addition, the advantage of using this technique is not only focused in reducing the pin count and the number of CLBs, but also in other types of radiation effects such as total ionization dose, as this method has the important characteristic of detecting permanent faults. So far, we have mentioned only SEUs that happen in the SRAM programmable cells that are permanent until the next reconfiguration. However, a circuit operating in the space environment can suffer from total ionization dose and other effects that can provoke permanent physical damages in the circuit.

Because there are many CED techniques, the next section evaluates the main CED techniques used in ASICs to detect a permanent effect of a SEU in arithmetic-based circuits synthesized in an SRAM-based FPGA. The goal is to investigate each one in terms of fault detection, area and performance penalties and to select the most appropriated ones for each type of circuit.

9.1.1 Using CED based on hardware redundancy

CED techniques based on hardware redundancy use extra hardware to compute the operation twice and compare the results. A direct way to implement it is the use of duplication with comparison (DWC) that simply duplicates the original hardware with the same operands and compares the results. The fault coverage depends on the observability of the fault by the input vectors. For single faults affecting only one of the circuits that compose the DWC scheme, there will be at least one input vector able to manifest the fault in the output. This technique has an area overhead of about 100% but almost no performance penalties, consequently, it is too costly and it will not be used to protect combinational circuits in the DWC-CED technique in SRAM-based FPGAs.

Another approach for CED based on hardware redundancy is to use extra hardware to compute different operands that are coded versions of the original ones, preferably with fewer bits, to minimize the area overhead. Any code can be used, but it is more appropriate to use a code that maintains the arithmetical and logical properties of the operands, to avoid the need of designing a totally new hardware to predict the new output. It means that, given two operands *a* and *b*, an operation *op* and a code *c*, the following equation must be valid:

$$c(a) \text{ op } c(b) = c(a \text{ op } b)$$

An interesting code with arithmetical properties is the residue code, also called module code. Residue code is applied by a recomputation of the remainders of the division of the operands by a given number. Figure 9-5 presents the schematic of a circuit using residue code as CED technique and figure 9-6 presents the implementation in VHDL.

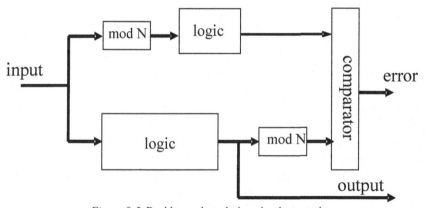

Figure 9-5. Residue code technique implementation

The version implemented in this paper uses module-15. The outputs of this code will have a maximum length of 4 bits. Most circuits used in this work have operands with 8 or more bits, so this code will surely provoke a reduction in the number of bits of the operands. The following piece of VHDL code shows the developed algorithm to calculate module 2n-1, for n=4 (module-15).

```
s <= ('1'&a(7 downto 4))-(not a(3 downto 0));
if (sub(4) = '1') then
moda := sub(3 downto 0); else
moda := a(7 downto 4) + a(3 downto 0); end if;
if (moda = "1111") then
mod_a := "0000"; end if;
```

Figure 9-6. Residue code technique implementation in VHDL

Because the work targets the investigation of techniques to detect the permanent effect of a SEU in SRAM-based FPGAs using the DWC-CED approach, the CED hardware redundancy based techniques are not attractive, because they can increase the area instead of reducing the costs. Residue code has less area overhead than DWC, depending on the width of the input of the original operands, so it can be used under some circumstances. However, it has performance penalties, depending on the delay of the residue encoder.

9.1.2 Using CED based on time redundancy

CED techniques based on time redundancy reduce the hardware cost at the expense of using extra time. It recomputes the operation in a different way to allow errors to be detected. During the first computation step, the normal operands are applied. In the recomputation step, the operands are encoded and a correct result can be generated after decoding. The mismatch of the two results indicates an error and, consequently, the presence of a fault in the circuit. In applications where performance is not essential, time redundancy is used to minimize the cost of the circuit, without increase in the circuit area or power consumption.

A very intuitive technique to use, but with limited applications, is the recomputing with swapped operands (RESWO). It can only be used in commutative operations, like adders and multipliers. It cannot be used for instance in division or subtraction operations. The RESWO technique tries to detect errors alternating the position of the operands. For example, after the computing of $a+b$, the operation $b+a$ can be done and the results compared to see if it is the same. Of course, it will not detect any faults if the two operands are equal, but it can have a high error detection capability in the other cases.

Another possible encoding technique is to use the distributive property of arithmetic logic to be able to identify faults. If one performs a 1-bit left shift of the input operands, it results in a multiplication by 2 of the operand. According to the operation, the result will be multiplied by 2 (adders) or by

4 (multipliers) and it can be easily divided by performing a 1-bit or 2-bit right shift in the output. This technique is called recomputing with shifted operands (RESO). Thus, in the first computation, the operands are computed and stored in a register. At the second computation, the operands are shifted k bits to the left, computed and the result is shifted k bits to the right (2k bits, if a multiplier or divider). In the proposed application, the operands are shifted by 1 bit. The result of the second step is compared to the previous result stored in the register. A mismatch indicates the presence of a permanent fault in the circuit. For example, in an adder, the left shifted operands are equal to the original ones multiplied by 2. The result of the sum should be the original result multiplied by 2 too. Then it is only necessary to shift right the new result and compare with the original one to detect a fault. The adder should be wide enough to add the shifted numbers without causing overflow. If not, a non existent fault can be wrong detected. Studies show the RESO detection capability (Patel; Fung, 1982).

For functions with operands of 8 bits, two approaches can be used: the use of the RESO with the same number of bits (8), or RESO with one more bit, to decrease the number of false detected faults. Of course, the second approach will result in an area overhead due to the new width of the operation.

Another option to increase the fault coverage with RESO is the use of one more clock cycle. Originally in the first cycle, the original operands are computed and in the next cycle, the left shifted operands are processed. If the fault is not detected yet, another clock cycle can be used, with the operands left shifted one more time (multiplied by 4 in the total). RESO increases the fault coverage as more shifts are applied to the operands (PATEL; FUNG, 1982). This approach will be called as 2-shift RESO, while the original approach as 1-shift RESO or only RESO. In order to increase the coverage, the performance will be depreciated due to the extra clock cycles. The schematic of a circuit using RESO is presented in figure 9-7.

Another option for time redundancy is the use of residue code, already presented in the hardware redundancy section. As the encoded operands have fewer bits than the original ones, the same hardware can be used to perform the original and the coded operands at two different moments. In order to use the same hardware, zeros must fill the non-used bits. The fault coverage will be reduced compared with the implementation using distinct hardware (one for the logic and the other for the encoding) because now both of the results are computed in the same faulty hardware.

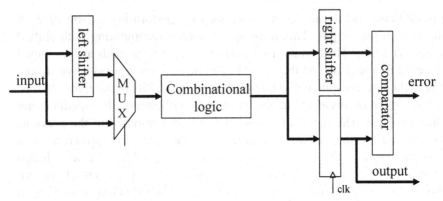

Figure 9-7. RESO technique implementation

In order to increase the fault coverage, these encoding techniques can be combined, one per each clock cycle. In the first clock cycle, the original operands are computed; in the second cycle, the operands using one type of encoding; and finally in the third cycle, the operands using another type of encoding. Of course, the drawback of this solution is the increase of area due to the use of two encoders and decoders, and the extra performance penalty with one more extra clock cycle.

9.1.3 Choosing the most appropriated CED block

In order to evaluate the fault coverage of the techniques previously presented, some combinational and sequential circuits were tested, including an 8-bit multiplier, arithmetic and logic unit (ALU), and a FIR canonical filter. Two tools, called Lemon Dragon multiplier and filter generator, automatically generated the multipliers and filters respectively. The tool provides two different syntheses: full array multipliers and constant array multipliers. Basically, several multipliers, adders and registers compose one FIR filter. To accomplish the goal of this paper, an automatic generation of fault injection structures was developed. All nodes in the design will be connected to exactly one fault injection component, so that the user may insert as many faults as needed. The components are described in VHDL language.

9.1.3.1 Multipliers

An 8-bit multiplier was the first case study. All techniques presented were implemented on this circuit: residue code with hardware redundancy; RESWO, 1-shift RESO with 8 bits (ignoring the left bit) and 9 bits (expanding the operands), 2-shift RESO with 8 and 9 bits, and residue code with time redundancy. The multipliers were implemented using cascaded full adders (FA), as shown in figure 9-8. For the 8-bit multiplier, there are 528 nodes, 1056 total faults (stuck-at 0 or 1), and for the 9-bit multiplier, 675 nodes, 1350 faults in total. In both cases, the two original operands have 8 bits, resulting in 2^{16} (65,536) combinations of input vectors. All combinations of faults and input vectors were injected, totaling 69,206,016 for the 8-bit version and 88,473,600 for 9-bit one.

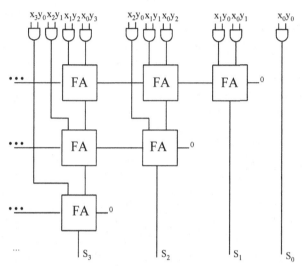

Figure 9-8. Multiplier using cascaded full adders

9.1.3.2 Arithmetic and Logic Unit (ALU)

The next case study was an Arithmetic and Logic Unit (ALU). This ALU performs the following operations: addition, subtraction, increment, decrement, AND, OR, XOR and NOT. It was designed in a bit slice approach, and the slice schematic is presented in figure 9-9. The operation is selected by signals $c1$, $c2$ and $c3$, operands are $a(i)$ and $b(i)$, cin is the carry in from the previous slice, the signal *cout* is the carry out to the next slice, and $s(i)$ is the output of the slice. This ALU has two input operands of 8 bits, plus 4 bits to select the operation. Then, there are 2^{20} (1,048,576) combinations of input vectors to be tested. Each slice has 16 nodes, resulting in 256 different faults for an 8-bit ALU. All the combinations of input vectors were tested with injected faults, totaling 268,435,456. At this time not all techniques were evaluated. The method RESWO was not used because some operations performed by the ALU are not commutative, like subtraction, increment or decrement.

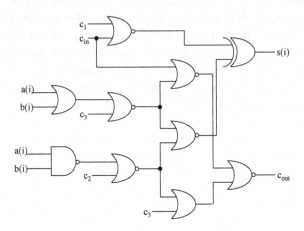

Figure 9-9. ALU bit slice

9.1.3.3 Digital FIR Filter

A canonical FIR filter circuit was chosen as a sequential case study for the proposed technique. Figure 8-10 showed the partial scheme of a canonical filter. The multipliers were designed with constant coefficients, resulting in an optimized area and minimal number of nodes. An 8-bit FIR canonical filter of 9 taps was automatically generated. The multiplier coefficients are: 2, 6, 17, 32 and 38. There is an 8-bit input; consequently, there are 2^8 (256) combinations of input vectors to test. The total of nodes in the FIR filter, including all its multipliers and adders is 4208. All the possible combinations of input vectors and faults were tested, totaling 1,077,248 runs.

9.1.4 Fault Coverage Results

The proposed DWC-CED technique for permanent fault detection was first validated by fault injection methodology in a prototype board using emulation. The fault injection system described in VHDL was specifically developed to test the proposed technique. Results were emulated in an AFX-PQ249-110 board using a XCV300 part. Some area comparisons between the proposed approach and TMR were also performed using Xilinx implementation tools.

The fault injection system is able to randomly choose the instant of insertion of the fault, the node and the redundant module (mult_dr0 or mult_dr1). There is a reset fault signal that works as a scrubbing, cleaning up the fault. Fault injection results show the reliability of the presented method. Figure 9-10 shows two graphics representing two different fault situations.

In the graph on the left , the fault occurs in module dr0 (st_perm_dr1=0, indicating that dr0 is fault free, consequently, trv2 receives dr1 (mux_select=1). Note that the value stored in the user's TMR register is held for one cycle, while the scheme identifies the free faulty module. In the graph the right side, a fault occurs in the module dr1 (st_perm_dr0=0, indicating that dr0 is fault free), as the default is register trv2 receiving dr0, nothing changes after the permanent fault detection.

For the exhaustive fault coverage evaluation, the following experiment was built based on the DWC-CED technique explained in section 9.2. Four versions of each case study circuit running in parallel were described in VHDL:

- Gold circuit used to compute the expected output of the circuits (module dr0)

- Copy of module dr0 with recomputing CED technique.

- Circuit under test (DUT) with fault injection capability, where the faults are injected (module dr1).

- Copy of module dr1 with recomputing CED technique, where the same faults are injected.

Note that in the real operation, the same hardware is used as DUT and for recomputation, consequently, there are only two modules: module dr0 and module dr1. However, for the experiment, two circuits for each redundancy module were used to perform both operations in parallel to reduce processing time. In addition, a prototype board (AFX-PQ240) was used to perform the fault injection experiment to speed up the fault emulation process.

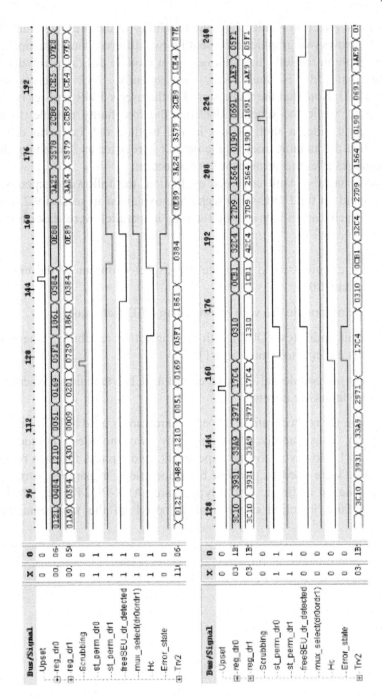

Figure 9-10. Upsets emulation in the Chipscope Analyzer (XILINX, 2001b) using the
Virtex® FPGA prototype board

In order to insert faults in all nodes of the case study circuits, a 4 to 1 multiplexor was inserted in each node in the VHDL description. If the select signal of the multiplexor is 00, the original signal is passed to the output; if select is 01, the constant 0 is the output (stuck-at 0); if select is 10, the constant 1 is propagated (stuck-at 1). The fault injection system operates with two clocks, one to control the change of the input vectors and other one to control the change of the faults. A counter controls the total number of combinations of input vectors and faults that must be inserted in the circuit. All combinations have been injected. There is a signal to indicate when the fault injection is completed.

In all cycles, the outputs of the gold circuit (module dr0) and the DUT (module dr1) are compared. If the outputs are equal (Hc = 0), this means that if there is a fault in one of the circuits, the fault did not generate an error in the output, so for real time operation proposes, this fault can be ignored and no detection operation must be performed. If a fault has generated an error in the output (Hc = 1), the output of module dr1 is compared with the decoded output of the recomputing circuit (copy of module dr1). If the outputs are not equal (Tc1 = 1), this means that the technique currently used was able to detect the fault. At the same time, the output of module dr0 is compared to the decoded output of the recomputing circuit (copy of module dr0). If the outputs are equal (Tc0 = 0), this means that the technique was able to detect a fault-free module.

An undetected fault is characterized when there is a mismatch in the output of dr0 and dr1 (Hc = 1) and the technique was not able to detect the faulty module (status Tc1 = 0) or it was not able to detect the fault-free module (status Tc0 = 1). A counter is incremented to show the number of total undetected faults. After all, this counter is read from the prototyped board and the percentage of undetected faults is calculated. The results in numbers and percentage of detected faults are in table 9-1.

Results show that all variations of RESO had better results in terms of fault coverage than residue code using time redundancy and RESWO. One can notice that residue code had higher fault coverage using hardware redundancy than time redundancy. This is because of the use of the same faulty hardware to compute the residue code, there is a high possibility of the coded word having the same effect at the output.

RESO is the most appropriate technique in terms of fault coverage for multipliers and consequently for all the circuits that use them, like filters. For the ALU, no one of the presented techniques was suitable enough to guarantee 100% of detection. This happens because the ALU logic is not only composed of arithmetic operations but also has logic Boolean functions, where the discussed techniques are not efficient.

9.1.4 Area and Performance Results

Table 9-2 presents area results of 8x8 and 16x16 bits multipliers, implemented in the XCV300 FPGA using no fault tolerance technique, TMR technique and the proposed technique (DWC-CED for permanent faults using RESO approach). All of the multipliers are synthesized with a register at the output. Results show that according to the size of the combinational logic block, it is possible to not only reduce the number of I/O pins but also area.

Table 9-1. Fault Coverage, Area and Performance Evaluation of CED techniques in SRAM-based FPGAs

Circuit	CED Technique	# of injected faults	# of detected faults	% of detected faults
8-bit Multiplier	Residue-15 (hard)	69,206,016	69,136,448	99.89
	Residue-15 (time)	69,206,016	47,387,924	68.47
	RESWO	69,206,016	48,458,171	70.02
	RESO 8 bits	69,206,016	69,176,011	99.95
	RESO 9 bits	88,473,600	88,473,600	100.00
	2-shift RESO 8 bits	69,206,016	69,198,150	99.98
8-bit ALU	Residue-15 (hard)	268,435,456	222,135,593	82.75
	Residue-15 (time)	268,435,456	199,912,813	74.47
	RESO 8 bits	268,435,456	213,005,264	79.35
	RESO 9 bits	268,435,456	245,694,848	91.52
	2-shift RESO 8 bits	268,435,456	213,048,871	79.36
	2-shift RESO 9 bits	268,435,456	245,763,385	91.55
	Residue-15+RESO-9bits	268,435,456	248,907,886	92.72
8-bit Filter	Residue-15 (hard)	1,077,248	1,077,248	100.00
	Residue-15 (time)	1,077,248	718,105	66.66
	RESO 8 bits	1,077,248	1,077,248	100.00

Note that the 16x16 bits multiplier protected by TMR could not be synthesized in the prototype board that uses a Virtex® part with 240 I/O pins (166 available for the user); while the same multiplier, implemented by the proposed technique could fit in the chip, and also occupy less area. In terms of performance, the TMR approach has presented a estimated frequency of 33.8 MHz, while the DMR-CED approach has presented a frequency of 26.7 MHz.

Table 9-2. Comparison of multiplier implementations (XCV300-PQ240)

Multipliers	Standard		TMR		DWC-CED	
Registered output	8x8	16x16	8x8	16x16*	8x8	16x16
Total of I/O pads	34	66	108	204	92 (-14%)	172 (-17%)
Number of 4-LUTs	159	741	584	2285	534 (-8,5%)	1791 (-22%)
Number of ffs	16	32	48	96	82 (+34)	162 (+66)
Non-registered output	8x8	16x16	8x8	16x16*	8x8	16x16
Total of I/O pads	32	64	96	192	66 (-31%)	130 (-32%)
Number of 4-LUTs	156	711	551	2159	425 (-23%)	1442 (-33%)
Number of ffs	0	0	0	0	34	66

* I/O pins were out of range, the part XCV300-BG432 was used.

As previously mentioned, according to the user's application requirements, the designer will be able to choose between a full hardware redundancy implementation (TMR) or a mixed solution, where time redundancy is combined with hardware redundancy to reduce pins and power dissipation in the interface. It is possible to use DMR and time redundancy only in the interface of the FPGA, in this way reducing pins. DMR and time redundancy can also be used in the design to reduce not only number of I/O pads, but also area for large combinational circuits as presented in table 9-2 and to increase reliability based on the concept published in (Lubaszewski; Courtois, 1998).

The same canonical FIR filter circuit presented in chapter 8 was used as a sequential case study circuit for the proposed technique, an 8-bit 9 taps filter with multiplier coefficients: 2, 6, 17, 32 and 38. The registers are also protected by TMR, while the combinational logic (multipliers and adders) is protected by DWC-CED using RESO approach. The CED block is placed at the outputs, and it votes the correct pad output from dr0 or dr1, as shown in figure 9-11.

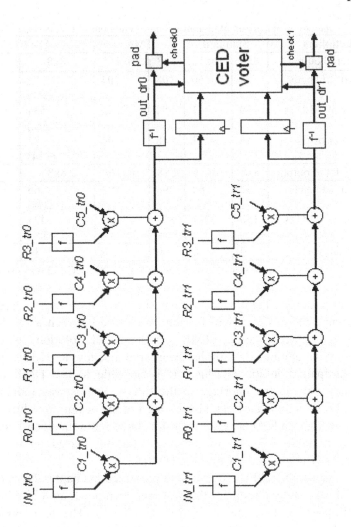

Figure 9-11. FIR Filter protected by DWC-CED technique

Table 9-3 presents area results of this filter using no tolerance technique, TMR technique and the proposed technique. Results show that the 9 taps FIR canonical filter occupies 13% less area in the FPGA if protected by DWC and time redundancy instead of by TMR. The results also present a reduction of 19% in the pin count compared to TMR. In terms of performance, the TMR has presented an estimated frequency of 40 MHz, while the DWC-CED technique has presented a frequency of 22 MHz. More results can be found in (Lima, Carro, Reis, 2003b).

Table 9-3. Filter Implementations XCV300-PQ240

	Standard	TMR	The proposed method
Total of I/O pads	26	84	68
Number of 4-LUTs	244	887	776
Number of ffs	64	192	226

In the case of the FIR digital filter, the technique can be additionally improved by using duplication in the registers too. The possibility of using duplication instead of TMR in the sequential logic is due to some characteristics of the filter. The first one is because the data inside the filter is pipelined. At each clock cycle, each register receives a new input that cleans up the upset that is propagated to the next register. In the worst case, it is necessary to wait the n clock cycles of the pipeline to wash out all the upsets. The second characteristic is the use of multiplier coefficients that are multiplied by a constant that usually corresponds to the highest possible input number to avoid floating point multiplications. This implies that the output must be divided by this same constant number, consequently the output is truncated and many upsets in the internal operation are eliminated in the end.

The test case is an 11 taps 9-bit digital low-pass filter protected by only DWC-CED in the combinational and sequential logic. The original fractional coefficients were multiplied by the constant 512. The final multiplier coefficients are: 1, -1, -9, 6, 73 and 120. There are ten 9-bit registers, totaling 90 bits that can be upset by SEU. Figure 9-13 shows some fault sensitive areas in the filter. An upset can affect the registers, which has a transient effect, or can affect the logic (multipliers, adders, voters), which has a permanent effect.

Based on the percentage of each type of memory cell in the whole set of memory elements in the CLBs, the LUTs represent 7.4%, the flip-flops represent 0.46%, the customization bits in the CLB represent 6.36% and the general routing represents 82.9%, the probability of an upset affecting the registers is very low compared to the probability of this same upset affecting the logic. In addition, the effect of an upset in a register is not always seen in the final output after being divided by the constant, in the example, the number 512.

Figure 9-13 shows the amplitude waveform of the input signal used in the case study filter. Figure 9-14 shows the amplitude waveform of the output of the filter in time domain. The input waveform has the frequencies 100Hz, 1KHz and 8KHz added in the same signal. The frequencies lower than 3.75KHz are passed to the output without any attenuation, in the example: frequencies 100Hz and 1KHz. Frequencies from 3.75 to 5.625 KHz are attenuated. Frequencies higher than 5.625 are blocked by the filter design.

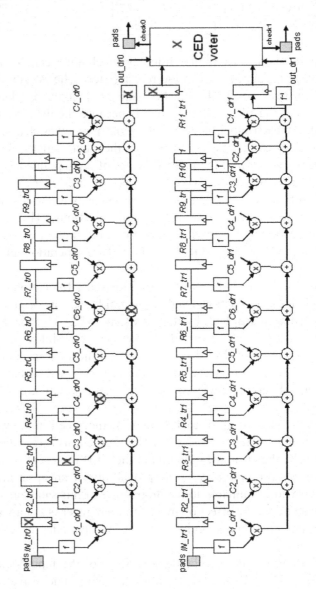

Figure 9-12. FIR Filter protected by DWC-CED technique in the combinational and sequential logic

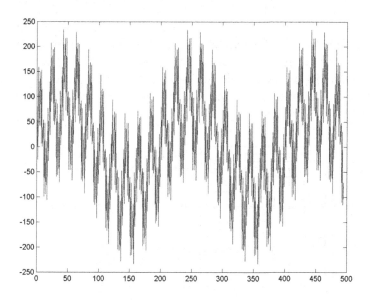

Figure 9-13. Amplitude signal input in the FIR filter

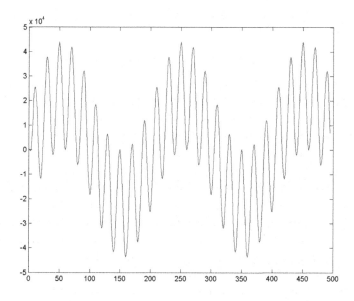

Figure 9-14. Amplitude signal output in the FIR filter

All possible combinations of bit flips for the tested input signal were injected in the registers. In total, 90 bit-flip faults were injected. Figure 9-15 shows the map of the bits in the filter. There are 9 bits multiplied by 10 registers, the fault bits from the first register need 10 clock cycles to be washed out, the fault bits from the second register need 9 clock cycles, the fault bits in the third register need 8 clock cycles, and so on. The calculation of the total number of clock cycles needed for the fault injection test is show in equation (1). Consequently, the filter is operating with the presence of faults for 495 clock cycles.

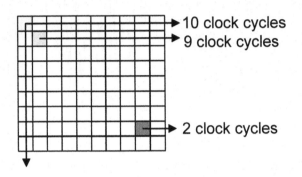

```
# clock cycles = 9x10 + 9x9 + 9x8 + 9x7 + 9x6 + 9x5 +
9x4 + 9x3 + 9x2 + 9x1   (1)
# clock cycles = 9x55 = 495
```

Figure 9-15. Map of the memory cells in the filter (9 bits x 10 registers)

Table 9-4 shows the effect of these upsets in the filter output. Note that less than 50% of the injected faults presented an effect in the 9 most significant bits of the output. Figure 9-16 shows the amplitude waveform of the output when faults were injected in the filter. Note that the signal has some noise compared to the original output.

In order to improve the integrity of the filter output signal, the 7 first tap registers, which are the ones that influence the most the output, had the 3 most significant bits (msb) protected by TMR, including the signal bit. In summary, 21 bits were protected from the total of 90, which represent 23% of the total sensitive bits. This protection reduces to upset effects in the output to a very low level as seen in graphics from the Matlab tool (Mathworks, 2003). Figure 9-17 shows the amplitude waveform of the filter output signal, in the presence of upsets, with the 21 bits protected by TMR. Note that the noise has been reduced to a very low level. Figures 9-18, 9-19 and 9-20 show the equivalent output signals in the frequency domain.

Table 9-4. The influence of the upsets injected in the registers in the filter output

Total number of injected bit flips	90
Total number of clock cycles in the presence of fault	495
Number of faulty clock cycles	487
Number of output faults in the 9 most significant bits	201
Number of output faults in the bit 9	43
Number of output faults in the bit 10	40
Number of output faults in the bit 11	31
Number of output faults in the bit 12	21
Number of output faults in the bit 13	21
Number of output faults in the bit 14	12
Number of output faults in the bit 15	13
Number of output faults in the bit 16	0
Number of output faults in the bit 17 (signal)	20

Table 9-5 shows a comparison between many SEU hardened filter implementations: the standard version, the TMR version, the filter protected by DWC-CED technique only in the combinational logic, the DWC-CED technique applied in the combinational and sequential logic, and the proposed DWC-CED technique applied in the combinational and in some bits of the sequential logic to improve reliability reducing cost.

Table 9-5. Filter Implementation using DWC-CED in the combinational and sequential logic (XCV300-PQ240)

	Standard	TMR	DWC-CED (combinational)	DWC-CED (all)	DWC-CED (*)
I/O pads	28	84	66	56	56
# 4-LUTs	496	1548	1350	1274	1304
# ffs	90	270	308	218	248

* 3 bits in the 7 first registers protected by TMR

Results show that for the 11 taps 9-bit FIR canonical filter protected by DMR and efficient TMR in only some bits of the registers occupies 3.5% less area in the FPGA compared with the DMR in the combinational logic and TMR in registers with 60 less flip-flops. Comparing with the full TMR, this new method shows a reduction of 16.5% in area and 22 less flip-flops.

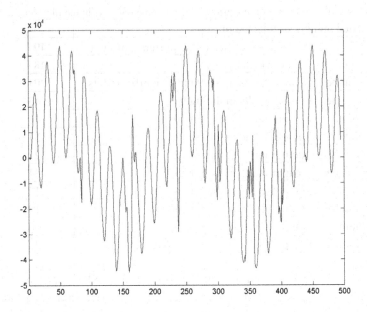

Figure 9-16. Amplitude signal output in the faulty FIR filter

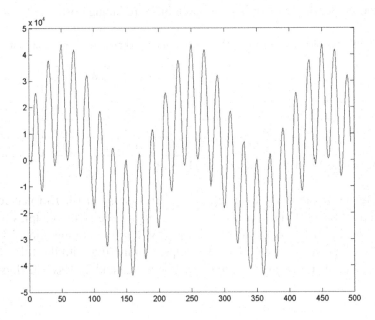

Figure 9-17. Amplitude signal output in the faulty FIR filter with 3-bit protected in the first 7 registers taps

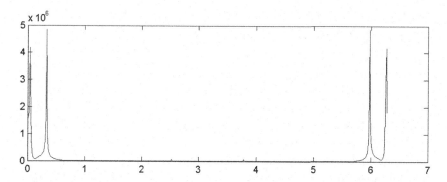

Figure 9-18. Signal output in the FIR filter in the frequency domain

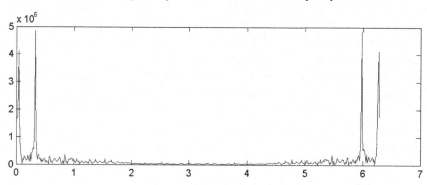

Figure 9-19. Signal output in the faulty FIR filter in the frequency domain

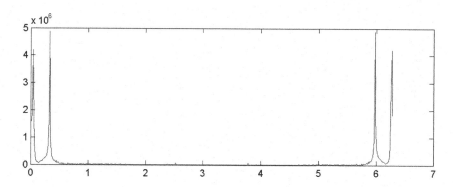

Figure 9-20. Signal output in the faulty FIR filter with 3-bit protected in the first 7 registers taps in the frequency domain

9.2 DESIGNING DWC-CED TECHNIQUE IN NON-ARITHMETIC-BASED CIRCUITS

The techniques previously presented are suitable for arithmetic-based circuits because they use some properties of the operation, but they are not convenient for random logic. An example of concurrent error detection for non-arithmetic based circuits is the parity prediction. The even/odd parity function indicates whether the number of 1's in a set of binary digits is even or odd. Techniques for designing datapath logic circuits and general combinational circuits with parity prediction have been described in (Nicolaidis; Duarte, 1998; Nicolaidis, 2003; Mitra; Mccluskey, 2002).

Figure 9-21 shows the basic architecture of a system with concurrent error detection using a single parity bit prediction. The circuit has n outputs and is designed in such a way that there is no sharing among the logic cones generating each of the outputs. Thus, a single fault can affect at most one output. The restriction of no logic sharing among different logic cones can result in large area overhead for circuits with a single parity bit prediction.

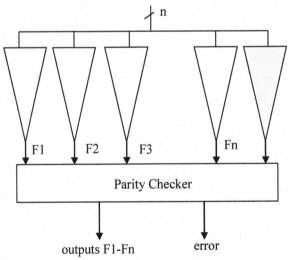

Figure 9-21. Parity prediction using single parity bit

Hence, the idea of using a single parity bit has been extended to multiple parity bits. This technique partitions the primary outputs into different parity groups. Sharing is allowed only among logic cones of the outputs that belong to different parity groups. There is a parity bit associated with the outputs in each parity group. The outputs of each parity group are checked using a parity checker. Figure 9-22 shows the general structure of a combinational logic circuit with two parity groups bit position. The parity of the outputs is predicted independently. The parity checker checks whether the actual parity of the outputs matches the predicted parity.

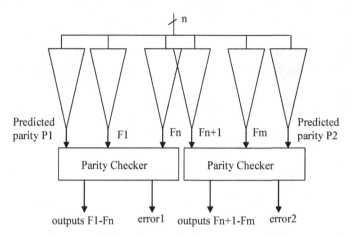

Figure 9-22. Multiple parity bits for concurrent error detection

The problem of using CED implemented by parity bit prediction is that many times the area occupied by the parity prediction logic is more than half of the original logic. Consequently, the final area result of the DWC-CED technique implemented with parity bit prediction can exceed the size of the TMR. But the advantage is still reduction in the number of input and output pads and possible increase in reliability (duplication with CED blocks).

Another example of CED for non-based arithmetic circuits is a technique based on unidirectional error detecting codes (Mitra; Mccluskey, 2002). A unidirectional error detecting code assumes that all errors are unidirectional; i.e., they change 0s to 1s or 1s to 0s but never both at the same time. Two unidirectional error detecting codes used for concurrent error detection are Berger codes and Bose-Lin codes. For the Berger code, a code-word is formed by appending a binary string representing the number of 0s (or the bit-wise complement of the number of 1s) to the given information word. Thus, for an information word consisting of n bits, the Berger code requires $n\log 2n$ extra bits to represent the number of 0s (or the bit-wise complement of number of 1s) in the information word. The Berger code has the capability of detecting all unidirectional errors. Figure 2-23 shows a concurrent error detection technique using Berger codes. Since the Berger code is a unidirectional error detection code, it is important to ensure that a single fault causes unidirectional errors at the outputs. This imposes a restriction that the logic circuits should be synthesized in such a way that they are inverter-free. Inverters can only appear at the primary inputs. In general, for Berger codes used to detect unidirectional errors on communication channels, the check-bits represent the bitwise complement of the number of 1's in the information word. However, since concurrent error detection techniques are designed to guarantee data integrity in the presence of single faults, a single fault can affect either the actual logic function or the logic circuit that predicts the number of 1's at the output but never both at the same time (since there is no logic sharing between the actual circuit and the circuit that predicts the number of 1's).

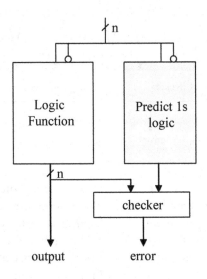

Figure 9-23. Unidirectional error detecting codes

The main conclusions presented in (Mitra; Mccluskey, 2002) show that results on benchmark circuits reveal marginal reduction in logic area by using CED schemes based on parity prediction instead of duplication. CED schemes based on Berger codes and Bose-Lin codes incur very high logic area overhead. It has been seen that it is important to analyze the properties of the combinational logic in order to choose the best technique in terms of fault coverage and area overhead. As future work, other solutions besides parity prediction and unidirectional error detecting codes, such as using prediction based on reversible logic function, will be also investigated to apply the DWC-CED method for non-based arithmetic logic.

Chapter 10

FINAL REMARKS

This book showed the study and development of SEU mitigation techniques for programmable architectures such as SRAM-based FPGAs. The choice of SRAM based FPGAs is due to their high applicability in apace applications. Because they are reprogrammable, designs can be updated or corrected after launch, which can reduce considerably the mission cost. The Virtex® family from Xilinx was chosen to be the case study for this work because is one of the most popular, highest logic density and best performing FPGAs in the market.

The problem of how to protect SRAM-based FPGAs at the architectural and at high-level methods was addressed in this book. Several fault-tolerant techniques able to protect integrated circuits against upsets in the combinational and sequential logic have been studied. The goal of this work was to investigate the techniques used nowadays and to develop new SEU mitigation techniques for SRAM-based FPGAs that are cost efficient in terms of time to market, low development cost, high performance, low area cost, low power dissipation and high reliability. In addition, FPGAs are becoming more complex with embedded hard microprocessors, such as the Virtex®II-Pro family from Xilinx. Consequently, the microprocessors must also be protected against upset.

In the first phase of the research, available techniques to protect integrated circuits against radiation were studied. The first case study circuit was the 8051 micro-controller from Intel. The microprocessor architecture was chosen for its representation of the majority of system requirements in space applications nowadays, presenting all types of logic to be protected and being part of the new generation architectures based on FPGA with an embedded hard microprocessor core. The description of the 8051 micro-controller used in the experiment was developed at UFRGS (Carro; Pereira;

Suzim, 1996). All registers and memories in the 8051 description were manually protected by hamming code (Lima et al., 2000; Lima et al., 2000b). A fault injection system built in VHDL was designed to test the protected version of the 8051 (Lima et al., 2001a). Results show a high reliability of the hamming code in presence of single upsets. The protected version was prototyped in a FPGA board from Altera and it has been tested under radiation ground test too. Results from the radiation show the necessity of using error correction code with multiple fault correction capability. In (Lima et al., 2002a), a fault injection study of the effect of multiple faults in the 8051 architecture is presented.

The second phase of the research has focused on the programmable field. A detailed analysis of the effect of a SEU in the programmable matrix of a SRAM-based FPGA was performed. When an upset occurs in the user's combinational logic implemented in a FPGA, it provokes a very peculiar effect not commonly seen in ASICs. The SEU behavior is characterized as a transient effect, followed by a permanent effect. The upset can affect either the combinational logic or the routing. The consequences of this type of effect, a transient followed by a permanent fault, cannot be handled by the standard fault tolerant solutions used in ASICs, such as Error Detection and Correction Codes (EDAC), Hamming code, or the standard TMR with a single voter, because a fault in the encoder or decoder logic or in the voter would invalidate the technique. The problem of protecting SRAM-based FPGAs against SEU is not well solved yet and more studies are required to improve the limitation of the methods currently used.

Some architectural solutions have been proposed along the book to improve reliability. One of them is the use of RS code combined with hamming code to protect the embedded memory against multiple upsets. This is an innovative solution that can be easily applied to any memory structure to protect against all double bit upsets and a large combination of multiple upsets. This technique was prototyped in a FPGA and results show that the area overhead is acceptable for the reliability achieved (Neuberger; Lima; Carro; Reis, 2003). One of the main advantages of this technique compared to the TMR is the low parity bits overhead, which in the case of the TMR is 200% and in the proposed approach varies around 10 to 20%. A drawback of this technique is the performance penalty. As future work, the encoder and decoder blocks will be implemented in ASIC to evaluate also the area and performance. We expect to get a lower area and performance penalty compared to the results from the FPGA prototype.

Another architectural proposed solution is based on the use of hardened memory cells with SET detection capability to replace the flip-flops located in the CLB in order to avoid bit flips and errors from transient faults in the combinational gates of the CLB, for instance, the multiplexors. This proposed approach can protect the flip-flop against SEU in the 1st, 2nd and 3rd order, and in addition to SET, which is a big concern in the very deep submicron technologies. As future work, a small prototype version of a SEU hardened FPGA protected by hardened memory cells and RS and hamming code will be designed (logic, simulation and layout) and tested in presence of faults.

However the main focus of this book is SEU mitigation techniques in high level description, which has been easily applied by the user with a low cost and a fast turnaround time for the market. Triple Modular Redundancy (TMR) with voters is a common high-level technique to protect ASICs against SEU and it can also be applied to protect FPGAs. The TMR technique was first tested in the Virtex® FPGA architecture by using a small design based on counters. Faults were injected in all sensitive parts of the FPGA by using the bitstream and a detailed analysis of the effect of a fault in a TMR design synthesized in the Virtex® platform was performed. This work has built a correlation between faults in the bitstream of the FPGA (one of the SRAM cells in the architecture) to the design logic synthesized in the FPGA. Results from fault injection and from radiation ground test facility showed the efficiency of the TMR for the related case study circuit.

In order to test a more complex design protected by TMR in the Virtex® platform that would also include embedded memories, the same 8051-like micro-controller description was protected by TMR and tested in the FPGA. The TMR 8051 micro-controller was tested by fault injection and under proton radiation in a ground facility. Fault injection analysis presented in (LIMA et al., 2001b) showed that there are a few upset bits in the bitstream related to the routing that can provoke an error in the TMR design. This limitation is due to the switch matrix that can connect two signals from different redundant parts when a programmable cell is upset. Based on the references presented in chapter 2, there is no totally efficient solution for SRAM based FPGAs that can ensure 100% of reliability in all conditions for SEU. This book had the goal of investigating the techniques used nowadays and to propose improvements in order to increase reliability. Although TMR has show high reliability, this technique presents some limitations, such as area overhead, three times more input and output pins and, consequently, a significant increase in power dissipation.

Aiming to reduce TMR costs and improving reliability, an innovative high-level technique for designing fault tolerant systems in SRAM-based FPGAs was developed, without modification to the FPGA architecture. The first proposed technique combines time and hardware redundancy to reduce area and pin count overhead (Lima, Carro, Reis, 2003a). This technique is based on duplication with comparison and time redundancy in the combination blocks of the design. It can be applied in arithmetic and in non-arithmetic circuits. Although the time redundancy technique can be successfully used to reduce pin count and area overhead over a full hardware redundancy, the transient concurrent error detection technique is not able to correct 100% of the faults occurring in FPGAs. Another penalty of this method is performance overhead because of the observation time. The evolution of this work investigates the use of modified time redundancy technique based on permanent fault detection to improve fault correction and to reduce the performance penalty at each clock cycle.

This technique was improved to a new one able to assure higher reliability with the same cost reduction. It is based on duplication with comparison and concurrent error detection (DWC-CED) (Lima, Carro, Reis, 2003b). This new technique proposed in this work was specifically developed for FPGAs to cope with transient faults that become permanent in the user combinational and sequential logic, while also reducing pin count, area and power dissipation. The RESO technique has been successfully applied in the DWC-CED approach proposed to detect and correct permanent faults in arithmetic circuits. The methodology was validated by fault injection experiments in an emulation board. Results in terms of area and pin count show reduction from 10 to 33% in the two cases studied (multipliers and digital filters). In addition, for digital filters, the DWC-CED approach can be applied in the combinational and sequential logic without loss in protection. For non-arithmetic based circuits, techniques such as parity prediction can be used in the DWC-CED method.

The technique DWC-CED has presented some performance penalties. As future work, improvements in this technique will be investigated to reduce the penalties in the performance. In addition, alternative techniques to detect permanent faults in non-arithmetic combinational circuits will be investigated. These techniques must present reduced area overhead and high fault coverage. According to the target application, it will be more important to reduce area, pin count or power dissipation. Another issue to be investigated is a technique to speed up the performance, increasing the area (there is always a compromise) for some specific applications, without reduce reliability.

The consideration of using FPGA in space applications is fairly recent and there is still a lot of work to be done in this area. In summary, the main contributions of this work were the detailed analysis of the effects of a single event upset (SEU) in the architecture of a SRAM-based FPGA, the investigation and experiment tests of the state-of-the-art fault-tolerant techniques and the development of a set of new SEU mitigation techniques that improve the reliability and reduce the cost in integrated circuits, and more specifically SRAM-based FPGAs.

REFERENCES

ACTEL INC. Using Synplify to Design in Actel Radiation-Hardened FPGAs: Application Report, USA, 2000. Available at: <www.actel.com/appnotes>. Visited on November, 2000.

ACTEL INC., RT54SX-S Rad-Tolerant FPGAs for Space Applications: Data Sheet. USA, 2001. Available at: <www.actel.com/datasheets>. Visited on August, 2001.

ALDERIGHI, M. et al. A Fault-Tolerant FPGA-based Multi-Stage Interconnection Network for Space Applications. In: IEEE INTERNATIONAL WORKSHOP ON ELECTRONIC DESIGN, TEST AND APPLICATIONS, DELTA, 1., 2002. Proceedings... IEEE Computer Society, 2002. p. 302-306.

ALEXANDRESCU, D.; ANGHEL, L.; NICOLAIDIS, M. New methods for evaluating the impact of single event transients in VDSM ICs. In: IEEE INTERNATIONAL SYMPOSIUM ON DEFECT AND FAULT TOLERANCE IN VLSI SYSTEMS WORKSHOP, DFT, 17., 2002. Proceedings... IEEE Computer Society, 2002. p. 99-107.

ALFKE, P.; PADOVANI, R. Radiation Tolerance on High-Density FPGAs. San Jose, USA: Xilinx, 1998.

ALTERA INC. Data Book. USA, 2001. Available at: <www.altera.com>. Visited on November, 2001.

ANGHEL, L.; ALEXANDRESCU, D.; NICOLAIDIS, M. Evaluation of a soft error tolerance technique based on time and/or space redundancy. In: SYMPOSIUM ON INTEGRATED CIRCUITS AND SYSTEMS DESIGN, SBCCI, 13., 2000. Proceedings... Los Alamitos: IEEE Computer Society, 2000. p. 237-242.

AVIZIENIS, A. Arithmetic Codes: Cost and Effectiveness Studies for Applications in Digital Systems Design. IEEE Transactions on Computer, New York, v.C-20, Nov. 1971.

ATMEL INC. Data Book. USA, 2001. Available at: <www.atmel.com>. Visited on November, 2001.

BARTH, J. Applying Computer Simulation Tools to Radiation Effects Problems. In: IEEE NUCLEAR SPACE RADIATION EFFECTS CONFERENCE, NSREC, 1997. Proceedings... IEEE Computer Society, 1997. p. 1-83.

BAUMANN, R.; SMITH, E. Neutron-induced boron fission as a major source of soft errors in deep submicron SRAM devices. In: IEEE INTERNATIONAL RELIABILITY PHYSICS SYMPOSIUM, 38., 2000. Proceedings... IEEE Computer Society, 2000.

BAUMANN, R. Soft errors in advanced semiconductor devices-part I: the three radiation sources. IEEE Transactions on Device and Materials Reliability, New York, v.1, n.1, p. 17-22, Mar. 2001.

BENS, H. et al. Low power radiation tolerant VLSI for advanced spacecraft. In: IEEE AEROSPACE CONFERENCE, 2002. Proceedings... IEEE Computer Society, 2002. p. 5-2401-5-2406.

175

BERNARDI, P.; REORDA, M. S.; STERPONE, L.; VIOLANTE, M. On the evaluation of SEUs sensitiveness in SRAM-based FPGAs, 10th IEEE International On-line Testing Symposium, 2004. pp. 115-120.

BESSOT, D.; VELAZCO, R. Design of SEU-hardened CMOS memory cells: the HIT Cell. In: EUROPEAN CONFERENCE ON RADIATION AND ITS EFFECTS ON COMPONENTS AND SYSTEMS, RADECS, 2., 1993. Proceedings... IEEE Computer Society, 1993. p. 563-570.

BETZ, V.; ROSE, J. FPGA Routing Architecture: Segmentation and Buffering to Optimize Speed and Density. In: ACM/SIGDA INTERNATIONAL SYMPOSIUM FIELD PROGRAMMABLE GATE ARRAY, FPGA, 1999. Proceedings... New York: ACM, 1999.

BOREL, J.; GAUTIER, J.; GASIOT, J. Silicon Redemption. In: EUROPEAN CONFERENCE ON RADIATION AND ITS EFFECTS ON COMPONENTS AND SYSTEMS, RADECS, 2001. Proceedings... IEEE Computer Society, 2001.

CAFFREY, M.; GRAHAM, P.; JOHNSON, E. Single Event Upset in SRAM FPGAs. In: MILITARY AND AEROSPACE APPLICATIONS OF PROGRAMMABLE LOGIC CONFERENCE, MAPLD, 2002. Proceedings..., 2002.

CALIN, T.; NICOLAIDIS, M.; VELAZCO, R. Upset hardened memory design for submicron CMOS technology. IEEE Transactions on Nuclear Science, New York, v.43, n.6, p. 2874 -2878, Dec. 1996.

CANARIS, J.; WHITAKER, S. Circuit techniques for the radiation environment of space. In: CUSTOM INTEGRATED CIRCUITS CONFERENCE, 1995. Proceedings... IEEE Computer Society, 1995. p. 77-80.

CARMICHAEL, C.; CAFFREY, M.; SALAZAR, A. Correcting Single-Event Upsets Through Virtex® Partial Configuration: Application Notes 216. San Jose, USA: Xilinx, 2000.

CARMICHAEL, C. Triple Module Redundancy Design Techniques for Virtex® Series FPGA: Application Notes 197. San Jose, USA: Xilinx, 2000.

CARMICHAEL, C.; FULLER, E.; FABULA, J.; LIMA, F. Proton Testing of SEU Mitigation Methods for the Virtex® FPGA. In: INTERNATIONAL CONFERENCE ON MILITARY AND AEROSPACE APPLICATIONS OF PROGRAMMABLE LOGIC DEVICES, MAPLD, 2001. Proceedings..., 2001.

CARRO, L.; PEREIRA, C.; SUZIM, A. Prototyping and reengineering of microcontroller-based systems. In: IEEE INTERNATIONAL WORKSHOP ON RAPID SYSTEM PROTOTYPING, RSP, 7., 1996. Proceedings... IEEE Computer Society, 1996. p. 178-182.

CARMICHAEL, C.; FULLER, E.; BLAIN, P.; CAFFREY, M. SEU Mitigation Techniques for Virtex® FPGAs in Space Applications. San Jose, USA: Xilinx, 1999.

COLINGE, J. Silicon-on-Insulator Technology: Overview and Device Physics. In: IEEE NUCLEAR SPACE RADIATION EFFECTS CONFERENCE, NSREC, 2001. Proceedings... IEEE Computer Society, 2001.

COTA, E.; LIMA, F.; REZGUI, S.; CARRO, L.; VELAZCO, R.; LUBASZEWSKI, M.; REIS, R. Synthesis of an 8051-like Micro-Controller Tolerant to Transient Faults. Journal of Electronic Testing Theory and Applications, JETTA, MA, USA, v.17, n.2, 2001.

COTA, E. et al. Implementing a self-testing 8051 microprocessor. In: SYMPOSIUM ON INTEGRATED CIRCUITS AND SYSTEMS DESIGN, SBCCI, 1999. Proceedings... Los Alamitos: IEEE Computer Society, 1999. p. 202-205.

CRAIN, S. et al. Analog and digital single-event effects experiments in space. IEEE Transactions on Nuclear Science, New York, v.48, n.6, Dec. 2001.

WHEN or Will FPGAs kill ASICs? Panel presented at ACM Design Automation Conference, DAC, 2001.

DELONG, T.A.; JOHNSON, B.W.; PROFETA, J. A. A fault injection technique for VHDL behavioral-level models. IEEE Design & Test of Computers, New York, v.13 n.4, p. 24-33, Winter 1996.

DENTAN, M. Radiation Effects On Electronic Components And Circuits. In: TRAINING COURSE OF THE EUROPEAN ORGANIZATION FOR NUCLEAR RESEARCH, CERN, 2000. Available at: <http://atlas.web.cern.ch/Atlas/GROUPS/FRONTEND /radhard.htm>. Visited on June, 2001.

DEPREITERE, J.; VAN MARCK, H.; VAN CAMPENHOUT, J. Evaluation of FPGA Switch Matrices using a Monte Carlo Approach. In: JAPANESE FPGA/PLD DESIGN CONFERENCE, 6., 1998. Proceedings..., 1998. p. 303-306.

DODD, P. E.; MASSENGILL, L. W. Basic Mechanism and Modeling of Single-Event Upset in Digital Microelectronics, IEEE Transaction on Nuclear Science, vol. 50, pp. 583-602, June 2003.

DUPONT, E.; NICOLAIDIS, M.; ROHR, P. Embedded robustness IPs for transient-error-free ICs. IEEE Design & Test of Computers, New York, v.19, n.3, p. 54-68, May-June 2002.

FULLER, E. et al. Radiation Testing Update, SEU Mitigation, and Availability Analysis of the Virtex® FPGA for Space Re-configurable Computing. In: IEEE NUCLEAR SPACE RADIATION EFFECTS CONFERENCE, NSREC, 2000. Proceedings... IEEE Computer Society, 2000.

FULLER, E. et al. Radiation test results of the Virtex® FPGA and ZBT SRAM for Space Based Reconfigurable Computing. In: INTERNATIONAL CONFERENCE ON MILITARY AND AEROSPACE APPLICATIONS OF PROGRAMMABLE LOGIC DEVICES, MAPLD, 2002. Proceedings..., 2002.

GAISLER, J. A portable and fault-tolerant microprocessor based on the SPARC v8 architecture. In: INTERNATIONAL CONFERENCE ON DEPENDABLE SYSTEMS AND NETWORKS, 2002. Proceedings... IEEE Computer Society, 2002. p. 409-415.

GUCCIONE, S. A.; LEVI, D.; SUNDARARAJAN, P. JBits: A Java-based interface for reconfigurable computing. In: Military and Aerospace Applications of Programmable Devices and Technologies Conference (MAPLD), Laurel, MD, Sept. 1999.

HASS, K. J.; TREECE, R. K.; GIDDINGS, A. E. A radiation-hardened 16/32-bit microprocessor. IEEE Transactions on Nuclear Science, New York, v.36, n.6, p. 2252-2257, Dec. 1989.

HASS, J. et al. Mitigating Single Event Upsets From Combinational Logic. In: NASA SYMPOSIUM ON VLSI DESIGN, 7., 1998. Proceedings..., 1998.

HASS, J. Probabilistic Estimates of Upset Caused by Single Event Transients. In: NASA SYMPOSIUM ON VLSI DESIGN, 8., 1999. Proceedings..., 1999.

HENTSCHKE, R.; MARQUES, F.; LIMA, F.; CARRO, L.; SUSIN, A.; REIS, R. Analyzing area and performance penalty of protecting different digital modules with hamming code and triple modular redundancy. In: SYMPOSIUM ON INTEGRATED CIRCUITS AND SYSTEMS DESIGN, 15., 2002. Proceedings..., Los Alamitos: IEEE Computer Society, 2002. p. 95-100.

HONEYWELL INC. MIL-STD-1705A Microprocessor Data Sheet. USA, 2003.

HOUGHTON, A. D. The Engineer's Error Coding Handbook. London: Chapman & Hall, 1997.

HUANG, W.; MCCLUSKEY, E. A. Memory Coherence Technique for Online Transient Error Recovery of FPGA Configurations. In: ACM/SIGDA INTERNATIONAL SYMPOSIUM ON FIELD PROGRAMMABLE GATE ARRAY, FPGA, 2001. Proceedings... New York: ACM, 2001. p. 183-192.

IBM INC. SOI Technology: IBM's Next Advance in Chip Design. USA, 2000. Available at: <www.ibm.com>. Visited on November, 2000.

INTEL INC. Embedded Micro-controllers Datasheet. USA, 1994.

IROM, F. et al. Single-event upset in commercial silicon-on-insulator PowerPC microprocessors. In: IEEE INTERNATIONAL SILICON-ON-INSULATOR CONFERENCE, 2002. Proceedings... IEEE Computer Society, 2002. p. 203-204.

JOHNSTON, A. Scaling and Technology Issues for Soft Error Rates. In: RESEARCH CONFERENCE ON RELIABILITY, 4., 2000. Proceedings... Palo Alto: Stanford University, 2000.

JOHNSON, B.; AYLOR, J. H.; HANA, H. Efficient Use of Time and Hardware Redundancy for Concurrent Error Detection in a 32-bit VLSI Adder. IEEE Journal of Solid-State-Circuits, New York, v.23, n.1, p. 208-215, Feb. 1988.

KASTENSMIDT, F. L.; STERPONE, L.; CARRO, L.; REORDA, M. On the Optimal Design of Triple Modular Redundancy Logic for SRAM-based FPGAs. In: Design, Automation and Test in Europe, DATE, 2005.

KATZ, R. et al. An SEU-Hard flip-Flop for Antifuse FPGAs. In: INTERNATIONAL CONFERENCE ON MILITARY AND AEROSPACE APPLICATIONS OF PROGRAMMABLE LOGIC DEVICES, MAPLD, 2001. Proceedings..., 2001.

KATZ, R. et al. Radiation effects on current field programmable technologies. IEEE Transactions on Nuclear Science, New York, v.44, n.6, p. 1945-1956, Dec. 1997.

KATZ, R. et al. Current radiation issues for programmable elements and devices. IEEE Transactions on Nuclear Science, New York, v.45, n.6, p. 2600-2610, Dec. 1998.

KATZ, R. et al. The effects of architecture and process on the hardness of programmable technologies. IEEE Transactions on Nuclear Science, New York, v.46, n.6, p. 1736-1743, Dec. 1999.

KINZEL, C., KASTENSMIDT, F. L., CARRO, L. Mapping the Virtex Customization Bits with JBits Classes for Selective Bitstream Fault Injection. IEEE Latin-American Test Workshop (6. : 2005 : Salvador, BA). [Digest of papers]. pp. 97-102.

KINZEL, C.; KASTENSMIDT, F. L.; CARRO, L. Improving Reliability of SRAM-Based Improving Reliability of SRAM-Based. In: European Conference On Radiation And Its Effects On Components And Systems (RADECS), 2005, New York : IEEE, 2005.

KUMAR, B. K. An FPGA Architecture with Error Correction Capability. In: ACM/SIGDA INTERNATIONAL SYMPOSIUM ON FIELD PROGRAMMABLE GATE ARRAYS, FPGA, 2003. Proceedings... New York: ACM, 2003.

LABEL, K. et al. A roadmap for NASA's radiation effects research in emerging microelectronics and photonics. In: IEEE AEROSPACE CONFERENCE, 2000. Proceedings... IEEE Computer Society, 2000. p. 535-545.

LACH, J.; MANGIONE-SMITH, W.; POTKONJAK, M. Efficient Error Detection, Localization and Correction for FPGA-Based Debugging. In: ACM/SIGDA INTERNATIONAL DESIGN AUTOMATION CONFERENCE, DAC, 2000. Proceedings... IEEE Computer Society, 2000.

LACH, J.; MANGIONE-SMITH, W.; POTKONJAK, M. Efficiently supporting fault-tolerance in FPGAs. In: ACM/SIGDA INTERNATIONAL SYMPOSIUM ON FIELD PROGRAMMABLE GATE ARRAYS, FPGA, 1998. Proceedings... New York: ACM, 1998. p. 105-115.

LEAVY, J. et al. Upset due to a single particle caused propagated transient in a bulk CMOS microprocessor. IEEE Transactions on Nuclear Science, New York, v.38, n.6, p. 1493-1499, Dec. 1991.

LERAY, J. Earth and Space Single-Events in Present and Future Electronics. In: EUROPEAN CONFERENCE ON RADIATION AND ITS EFFECTS ON COMPONENTS AND SYSTEMS, RADECS, 6., 2001. Short Course. IEEE Computer Society, 2001.

LIMA, F.; COTA, E.; CARRO, L.; LUBASZEWSKI, M.; REIS, R.; VELAZCO, R.; REZGUI, S. Designing a radiation hardened 8051-like micro-controller. In: SYMPOSIUM ON INTEGRATED CIRCUITS AND SYSTEMS DESIGN, SBCCI, 13., 2000. Proceedings... Los Alamitos: IEEE Computer Society, 2000a. p. 255-260.

LIMA, F.; REZGUI, S.; COTA, E.; CARRO, L.; LUBASZEWSKI, M.; VELAZCO, R.; REIS, R. Designing and Testing a Radiation Hardened 8051-like Micro-controller. In: INTERNATIONAL CONFERENCE ON MILITARY AND AEROSPACE APPLICATIONS OF PROGRAMMABLE LOGIC DEVICES, MAPLD, 2000. Proceedings..., 2000b.

LIMA, F.; REZGUI, S.; CARRO, L.; VELAZCO, R.; REIS, R. On the use of VHDL simulation and emulation to derive error rates. In: EUROPEAN CONFERENCE ON RADIATION AND ITS EFFECTS ON COMPONENTS AND SYSTEMS, RADECS, 2001. Proceedings... IEEE Computer Society, 2001a. p. 253-260.

LIMA, F.; CARMICHAEL, C.; FABULA, J.; PADOVANI, R.; REIS, R. A fault injection analysis of Virtex® FPGA TMR design methodology. In: EUROPEAN CONFERENCE ON RADIATION AND ITS EFFECTS ON COMPONENTS AND SYSTEMS, RADECS, 2001. Proceedings... IEEE Computer Society, 2001b. p. 275-282.

LIMA, F.; CARRO, L.; VELAZCO, R.; REIS, R. Injecting Multiple Upsets in a SEU tolerant 8051 Micro-controller. In: LATIN AMERICA TEST WORKSHOP, LATW, 2002. Proceedings... Amissville: IEEE Computer Society, 2002a.

LIMA, F.; CARRO, L.; VELAZCO, R.; REIS, R. Injecting multiple upsets in a SEU tolerant 8051 micro-controller. In: IEEE INTERNATIONAL ON-LINE TESTING WORKSHOP, IOLTW, 8., 2002. Proceedings... IEEE Computer Society, 2002b. p. 194.

LIMA, F.; CARRO, L.; REIS, R. Prototyping, verification, and test: Reducing pin and area overhead in fault-tolerant FPGA-based designs. In: ACM/SIGDA INTERNATIONAL SYMPOSIUM ON FIELD PROGRAMMABLE GATE ARRAYS, FPGA, 2002. Proceedings... New York: ACM, 2003a. p. 108-117.

LIMA, F.; CARRO, L; REIS, R. Techniques for reconfigurable logic applications: Designing fault tolerant systems into SRAM-based FPGAs. In: INTERNATIONAL DESIGN AUTOMATION CONFERENCE, DAC, 2003. Proceedings... New York: ACM, 2003b. p. 650-655.

LIU, M. N.; WHITAKER, S. Low power SEU immune CMOS memory circuits. IEEE Transactions on Nuclear Science, New York, v.39, n.6, p. 1679-1684, Dec. 1992.

LUBASZEWSKI, M.; COURTOIS, B. A reliable fail-safe system. IEEE Transactions on Computers, New York, v.47, n.2, p. 236-241, Feb. 1998.

LUM, G.; MARTIN, L. Single Event Effects Testing of Xilinx FPGAs. San Jose, USA: Xilinx, 1998.

MATHWORKS INC. Matlab and Simulink Documentation. USA, 2003.

MAVIS, D.; EATON, P. SEU and SET Mitigation Techniques for FPGA Circuit and Configuration Bit Storage Design. In: INTERNATIONAL CONFERENCE ON MILITARY AND AEROSPACE APPLICATIONS OF PROGRAMMABLE LOGIC DEVICES, MAPLD, 2000. Proceedings..., 2000.

MAVIS, D.; EATON, P. Soft error rate mitigation techniques for modern microcircuits. In: RELIABILITY PHYSICS SYMPOSIUM, 40., 2002. Proceedings..., 2002. p. 216-225.

MAVIS, D. et al. A Reconfigurable, Nonvolatile, Radiation Hardened Field Programmable Gate Array (FPGA) for Space Applications. In: INTERNATIONAL CONFERENCE ON MILITARY AND AEROSPACE APPLICATIONS OF PROGRAMMABLE LOGIC DEVICES, MAPLD, 1998. Proceedings..., 1998.

MAXWELL TECHNOLOGIES. Product Data sheet. USA, 2001. Available at: <www.spaceelectronics.com/>. Visited on January, 2002.

MESSENGER, G. C. Collection of Charge on Junction Nodes from Ion Tracks, IEEE Trans. Nuclear Science, vol. NS-29, pp. 2024-2031, Dec. 1982.

MITRA, S.; MCCLUSKEY, E. Which Concurrent Error Detection Scheme To Choose? In: INTERNATIONAL TEST CONFERENCE, ITC, 2002. Proceedings... IEEE Computer Society, 2002.

MITRA, S.; SHIRVANI, P.; MCCLUSKEY, E. Fault Location in FPGA-Based Reconfigurable Systems. In: WORKSHOP ON DEFECT AND FAULT-TOLERANCE IN VLSI SYSTEMS, 1998. Proceedings... IEEE Computer Society, 1998.

MOORE, G. E. Progress in Digital Integrated Electronics. Digest of the 1975 International Electron Devices Meeting, New York, p. 1113, 1975.

MUSSEAU, O.; FERLET-CAVROIS, V. Silicon-on-Insulator Technology: Radiation Effects. In: IEEE NUCLEAR SPACE RADIATION EFFECTS CONFERENCE, NSREC, 2001. Short Course. IEEE Computer Society, 2001.

NASA. Radiation Effects on Digital Systems. USA, 2002. Available at: <radhome.gsfc.nasa.gov/top.htm>. Visited on January, 2003.

NEUBERGER, G.; LIMA, F.; CARRO, L.; REIS, R. A Multiple Bit Upset Tolerant SRAM Memory. Transactions on Design Automation of Electronic Systems, TODAES, New York, v.8, n.4, Oct. 2003.

NICOLAIDIS, M.; PEREZ, R. Measuring the width of transient pulses induced by radiation. In: IEEE INTERNATIONAL RELIABILITY PHYSICS SYMPOSIUM, 2003. Proceedings... IEEE Computer Society, 2003. p. 56-59.

NICOLAIDIS, M. Carry checking/parity prediction adders and ALUs. IEEE Transactions on Very Large Scale Integration (VLSI) Systems, New York, v.11, n.1, p. 121-128, Feb. 2003.

NICOLAIDIS, M.; DUARTE, R. O. Design of fault-secure parity-prediction Booth multipliers. In: INTERNATIONAL CONFERENCE ON DESIGN, AUTOMATION AND TEST IN EUROPE, DATE, 1998. Proceedings... IEEE Computer Society, 1998. p. 7-14.

NICOLAIDIS, M. Time redundancy based soft-error tolerance to rescue nanometer technologies. In: IEEE VLSI TEST SYMPOSIUM, 17., 1999. Proceedings..., IEEE Computer Society, 1999. p. 86-94.

NORMAND, E. Correlation of in-flight neutron dosimeter and SEU measurements with atmospheric neutron model. IEEE Transactions on Nuclear Science, New York, v.48, n.6, p. 1996-2003, Dec. 2001.

NORMAND, E.; BAKER, T. J. Altitude and latitude variations in avionics SEU and atmospheric neutron flux. IEEE Transactions on Nuclear Science, New York, v.40, n.6, p. 1484-1490, Dec. 1993.

NORMAND, E. Single event upset at ground level. IEEE Transactions on Nuclear Science, New York, v.43, n.6, p. 2742-2750, Dec. 1996.

O'BRYAN, M., LABEL, K. Recent Radiation Damage and Single Event Effect Results for Candidate Spacecraft Electronics. In: IEEE NUCLEAR SPACE RADIATION EFFECTS CONFERENCE, NSREC, 2001. Proceedings... IEEE Computer Society, 2001.

O'BRYAN, M. et al. Current single event effects and radiation damage results for candidate spacecraft electronics. In: IEEE RADIATION EFFECTS DATA WORKSHOP, 2002. Proceedings... IEEE Computer Society, 2002. p. 82-105.

OHLSSON, M.; DYREKLEV, P.; JOHANSSON, K.; ALFKE, P. Neutron Single Event Upsets in SRAM based FPGAs. In: IEEE NUCLEAR SPACE RADIATION EFFECTS CONFERENCE, NSREC, 1998. Proceedings... IEEE Computer Society, 1998.

PATEL, J. H.; FUNG, L. Y. Concurrent Error Detection in ALUs by Recomputing with Shifted Operands. IEEE Transactions on Computer, New York, v.C-31, July 1982.

PATEL, J.; FUNG, L. Multiplier and Divider Arrays with Concurrent Error Detection. In: INTERNATIONAL SYMPOSIUM ON FAULT-TOLERANT COMPUTING, 1996. Proceedings... IEEE Computer Society, 1996.

PETERSON, W. Wesley. Error-correcting codes. 2nd. ed. Cambridge : The Mit Press, 1980. 560p.

REBAUDENGO, M.; REORDA, M. S.; VIOLANTE, M. Simulation-based Analysis of SEU effects of SRAM-based FPGAs. In: INTERNATIONAL WORKSHOP ON FIELD-PROGRAMMABLE LOGIC AND APPLICATIONS, FPL, 2002. Proceedings... IEEE Computer Society, 2002. p. 607-615.

REBAUDENGO, M.; REORDA, M. S.; VIOLANTE, M.; NICOLESCU, B.; VELAZCO, R. Coping with SEUs/SETs in microprocessors by means of low-cost solutions: a comparison study. IEEE Transactions on Nuclear Science, New York, v.49, n.3, June 2002.

REDINBO, G.; NAPOLITANO, L.; ANDALEON, D. Multi-bit Correcting Data Interface or Fault-Tolerant Systems. IEEE Transactions on Computers, New York, v.42, n.4, p. 433-446, Apr. 1993.

REED, R. A. et al. Heavy ion and proton-induced single event multiple upset. IEEE Transactions on Nuclear Science, New York, v.44, n.6, p. 2224-2229, Dec. 1997.

ROCKETT, L. R. A. design based on proven concepts of an SEU-immune CMOS configurable data cell for reprogrammable FPGAs. Microelectronics Journal, Elsevier, v.32, p. 99-111, 2001.

ROCKETT, L. R. An SEU-hardened CMOS data latch design. IEEE Transactions on Nuclear Science, New York, v.35, n.6, p. 1682-1687, Dec. 1988.

SEXTON, F. et al. SEU simulation and testing of resistor-hardened D-latches in the SA3300 microprocessor. IEEE Transactions on Nuclear Science, New York, v.38, n.6, p. 1521-1528, Dec. 1991.

SIA SEMICONDUCTOR INDUSTRY ASSOCIATION. The National Technology Roadmap for Semiconductors. USA, 1994.

SILVA, L.; LIMA, F.; CARRO, L.; REIS, R. Synthesis of the FPGA Version of 8051. In: UFRGS MICROELECTRONICS SEMINAR, 12., 1997, Porto Alegre. Proceedings... Porto Alegre: CPGCC da UFRGS, 1997. p. 115-120.

SKAHILL, K. VHDL for Programmable Logic. Addison Wesley. 1996. p. 1-23.

STASSINOPOULOS, E.; RAYMOND, J. The space radiation environment for electronics. Proceedings of the IEEE, New York, v.76, n.11, p. 1423-1442, Nov. 1988.

STURESSON, F.; MAUSSON, S.; CARMICHAEL, C.; HARBOE-SORENSEN, R. Heavy ion characterization of SEU mitigation methods for the Virtex® FPGA. In: EUROPEAN CONFERENCE ON RADIATION AND ITS EFFECTS ON COMPONENTS AND SYSTEMS, RADECS, 2001. Proceedings... IEEE Computer Society, 2001. p. 285-291.

SWIFT, G. M.; REZGUI, S.; GEORGE, J.; CARMICHAEL, C.; NAPIER, M.; MAKSYMOWICZ, J.; MOORE, J.; LESEA, A.; KOGA, R.; WROBEL, T. F. Dynamic testing of Xilinx Virtex®-II field programmable gate array (FPGA) input/output blocks (IOBs)., IEEE Transactions on Nuclear Science, v.51, n.6, p. 3469-3474, Dec. 2004.

VELAZCO, R. et al. Two CMOS memory cells suitable for the design of SEU-tolerant VLSI circuits. IEEE Transactions on Nuclear Science, New York, v.41, n.6, p. 2229-2234, Dec. 1994.

VELAZCO, R.; CHEYNET, P.; ECOFFET, R. Effects of radiation on digital architectures: one year results from a satellite experiment. In: SYMPOSIUM ON INTEGRATED CIRCUITS AND SYSTEMS DESIGN, SBCCI, 1999. Proceedings... Los Alamitos: IEEE Computer Society, 1999. p. 164-169.

VELAZCO, R.; REZGUI, S.; ECOFFET, R. Transient bitflip injection on microprocessor-based digital architectures. In: IEEE NUCLEAR AND SPACE RADIATION EFFECTS CONFERENCE, NSREC, 2000. Proceedings... IEEE Computer Society, 2000.

WANG, J.; WONG, W.; WOLDAY, S.; CRONQUIST, B.; MCCOLLUM, J., KATZ, R.; KLEYNER, I. Single Event Upset and Hardening in 0.15µm Antifuse-Based Field Programmable Gate Array, IEEE Transactions On Nuclear Science, v.50, n.6, p. 2158-2166, Dec. 2003.

WANG, J. et al. Clock buffer circuit soft errors in antifuse-based field programmable gate arrays. IEEE Transactions on Nuclear Science, New York, v.47, n.6, Dec. 2000.

WANG, J. et al. SRAM based re-programmable FPGA for space applications. IEEE Transactions on Nuclear Science, New York, v.46, n.6, p. 1728-1735, Dec. 1999.

WEAVER, H.; et al. An SEU Tolerant Memory Cell Derived from Fundamental Studies of SEU Mechanisms in SRAM. IEEE Transactions on Nuclear Science, New York, v.34, n.6, Dec. 1987.

WIRTH, G. I.; VIEIRA, M. G.; HENES NETO, Egas; KASTENSIMIDT, F. G. L. Single Event Transients in Combinatorial Circuits. 18th International Symposium on Integrated Circuits and Systms Design, (Florianópolis, Brazil). In: Proceedings ... New York (USA) : ACM - Association for Computing Machinery, 2005. p. 121-126, 2005a.

WIRTH, G. I.; VIEIRA, M. G.; KASTENSIMIDT, F. G. L. Computer Efficient Modeling of SRAM Cell Sensitivity to SEU. In: IEEE Latin American Test Workshop, 2005, Salvador - Bahia. IEEE Latin American Test Workshop. Amissville (USA): IEEE Computer Society, 2005, p. 51-56, 2005b.

WHITAKER, S.; CANARIS, J.; LIU, K. SEU hardened memory cells for a CCSDS Reed-Solomon encoder. IEEE Transactions on Nuclear Science, New York, v.38, n.6, p. 1471-1477, Dec. 1991.

WISEMAN, D. et al. Design and Testing of SEU / SEL Immune Memory and Logic Circuits in a Commercial CMOS Process. In: IEEE NUCLEAR SPACE RADIATION EFFECTS CONFERENCE, NSREC, 1993. Proceedings... IEEE Computer Society, 1993.

XILINX, INC. Virtex®™ 2.5V Field Programmable Gate Arrays: Datasheet DS003. USA, 2000a.

XILINX, INC. QPRO™Virtex®™ 2.5V Radiation Hardened FPGAs: Application Notes 151. USA, 2000b.

XILINX INC. Virtex® Series Configuration Architecture User Guide: Application Notes 151. USA, 2000c.

XILINX INC. Virtex®-II IP-Immersion™ Technology Enables Next-Generation Platform FPGAs. Xcell Journal Online. USA, 2001a.

XILINX INC. ISE Software and Chipscope Analyzer Manual. USA, 2001b.

XU, J. et al. A Novel Fault Tolerant Approach for SRAM-Based FPGAs. In: PACIFIC INTERNATIONAL SYMPOSIUM ON DEPENDABLE COMPUTING, 1999. Proceedings... IEEE Computer Society, 2000.

YU, Shu-Yi, MCCLUSKEY, E. Permanent Fault Repair for FPGAs with Limited Redundant Area. In: DESIGN FOR FAULT TOLERANCE CONFERENCE, DFT, 2001. Proceedings... IEEE Computer Society, 2001.

ZOUTENDYK, J.; EDMONDS, L.; SMITH, L. Characterization of multiple-bit errors from single-ion tracks in integrated circuits. IEEE Transactions on Nuclear Science, New York, v.36, n.6, p. 2267-2274, Dec. 1989 .